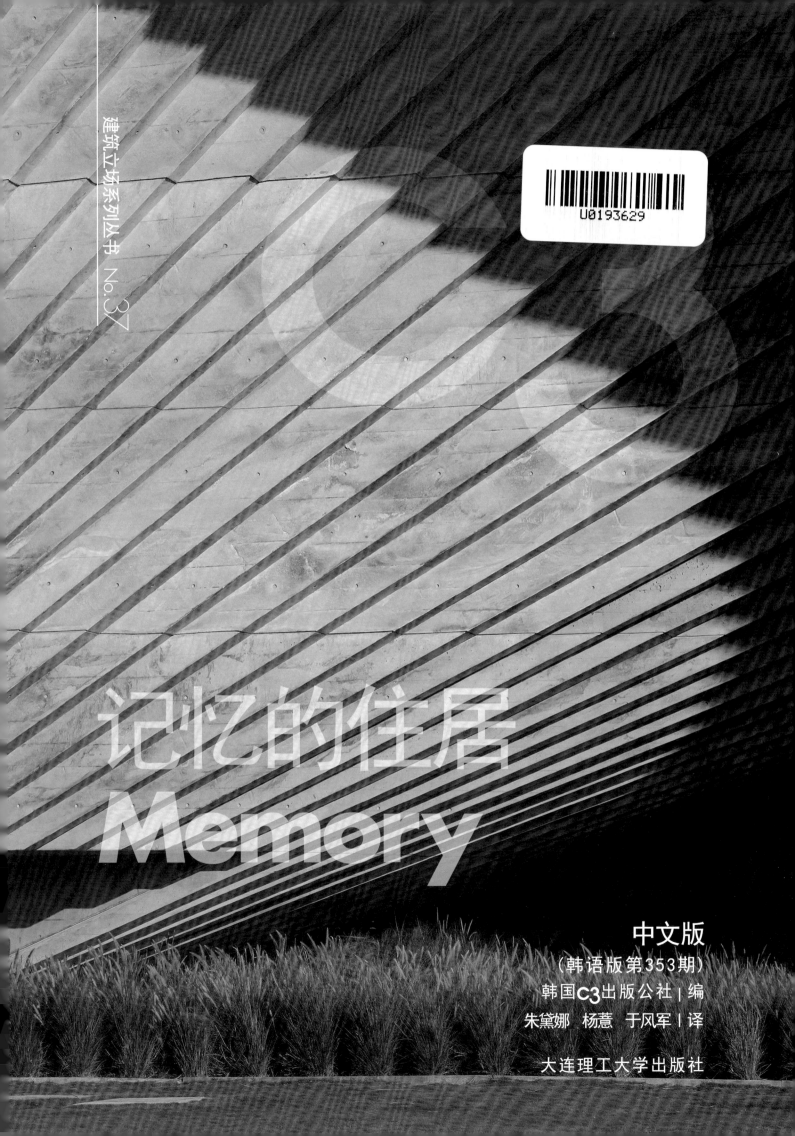

建筑立场系列丛书 No.37

记忆的住居
Memory

中文版
（韩语版第353期）
韩国C3出版公社 | 编
朱黛娜 杨薏 于风军 | 译

大连理工大学出版社

# 记忆的住居

- 004 情感记忆的表面 _ Paula Melâneo
- 008 菲希特尔贝尔格的木屋 _ AFF Architekten
- 018 C号房，一处修缮后的干草仓 _ Campovono Baumgartner Architekten
- 028 庄园马厩 _ AR Design Studio
- 036 花园树屋 _ Hironaka Ogawa & Associates

## 新世界——源于旧工厂场地

- 044 将旧工业用地改建为新世界 _ Tom Van Malderen
- 048 丹麦国家海洋博物馆 _ BIG
- 066 瓦尔帕莱索文化公园 _ HLPS Arquitectos
- 082 景观实验室 _ Cannata & Fernandes Arquitectos
- 094 Casa Mediterraneo总部 _ Manuel Ocaña
- 106 Conde Duque建筑 _ Carlos de Riaño Lozano
- 118 Can Ribas工厂的修复 _ Jaime J. Ferrer Forés

## 彰显自由与功能的
### 大学建筑类型学

- 134 新的知识纪念碑 _ Aldo Vanini
- 138 RGS艺术、建筑和设计中心 _ Tadao Ando Architect & Associates
- 150 圣豪尔赫大学卫生学院 _ Taller Básico de Arquitectura
- 158 利默里克大学医学院和学生公寓 _ Grafton Architects
- 168 维也纳对外经济贸易大学的法学院和行政大楼 _ CRAB Studio

- 182 建筑师索引

C3 建筑立场系列丛书 No.37

# Memory

004 *Surfaces of Affective Memories* _ Paula Melâneo

008 Hut in Fichtelberg _ AFF Architekten

018 House C, A Refurbished Hay Barn _ Campovono Baumgartner Architekten

028 Manor House Stables _ AR Design Studio

036 Garden Tree House _ Hironaka Ogawa & Associates

## New Reality from Old Industrial Site

044 *Turning Former Industrial Sites into New Realities* _ Tom Van Malderen

048 Danish National Maritime Museum _ BIG

066 Valparaíso Cultural Park _ HLPS Arquitectos

082 Landscape Laboratory _ Cannata & Fernandes Arquitectos

094 Casa Mediterraneo Headquarters _ Manuel Ocaña

106 Conde Duque _ Carlos de Riaño Lozano

118 Can Ribas Factory Renovation _ Jaime J. Ferrer Forés

## Liberal and Functional
### University Faculty Building Typology

134 *New Monuments to Knowledge* _ Aldo Vanini

138 Roberto Garza Sada Center for Arts, Architecture and Design
    _ Tadao Ando Architect & Associates

150 Health Faculty of San Jorge University _ Taller Básico de Arquitectura

158 Medical School, Student Residences at the University of Limerick _ Grafton Architects

168 Vienna University's Law and Administration Buildings _ CRAB Studio

182 Index

# 记忆的住居

从词源学角度上来看,记忆作为一项精神活动,是能够保留住想法和图像的一种能力。它是一个包含有多种关系的复杂体系,其自身在建筑方面的体现也是人类生活和文化的一部分。

有人将文学视为一种类推,而南美小说家Paul Auster则认为,"记忆是一件事情能发生两次的地方",从这种角度说来,与记忆相关的建筑工程可能代表建造一项蕴含有建筑师精心挑选并推介的一些指导方针内容的第二次机会。

与过往时代相关的一些极具特色的工程比较关注在一定物理空间和一群人内部,具有本土风格、历史或个性的恢复,然而,一个地方的特色通过对当地环境、人类的存在和生命足迹的理解——按当地的水平——以及对于目前生活风格来说开创一些新意从而得以保存。

修复美化和重复使用在每个项目中都具有非常特殊和明确的意义。现有的已修建好的元素没有保留相同的功能,材料也没有被再次使用——就像异教的建筑物的石料被用来建造天主教教堂——然而,这些居然是可行的,因为它们有一个与其表面相符的历史和内容。作为一张羊皮纸,这些提议成为改写每日历史的全新媒介。

In its etymological sense, memory is the faculty of retaining ideas and images, as a mental process. It consists of a complex system of relations which is part of human life and culture, with its own reflexes in architecture.

Someone brings literature as an analogy and the words of the North-American fiction writer Paul Auster *"Memory is the space in which a thing happens for a second time"*, by this extent, architecture projects dealing with memory might represent a second opportunity to construct a narrative, where some guidelines are deliberately chosen and introduced by the architects.

Related with the times past, the featured projects are concerned about retrieving part of a local atmosphere, history or identity, circumscribed in a physical space and within a group of people. Somehow the genius loci (the spirit of a place) is preserved here, by understanding the environment, the human existence and life roots – at a local level –, and by creating new meaningful places for contemporary lifestyle.

Rehabilitation and re-use acquire a very special and specific sense in each project. The existing constructed elements don't maintain the same functions and the materials are not re-used per se – just like stones from pagan constructions were used to build catholic churches – but they are worked-out because they have a history or a narrative attached to its surface. As a palimpsest, these proposals act as new mediums to re-write everyday histories.

菲希特尔贝尔格的木屋/AFF Architekten
C号房，一处修缮后的干草仓/Campovono Baumgartner Architekten
庄园马厩/AR Design Studio
花园树屋/Hironaka Ogawa & Associates

情感记忆的表面/Paula Melâneo

## 情感记忆的表面

记忆是历史和建筑研究中的核心因素。我们对于过往和自我意识的理解推动了历史学家和建筑师对于所重建的内容及方式的相关理论研究。对此，人们首创了一些用来保护和保存这些内容的严格制度，其中大部分是源于18—19世纪，且有些在今天依然是有效的。

这些想法促成了全球范围内人类历史一致的集体性记忆，但是它们有时也会脱离一些诸如本地传统、可持续性、生态学，特别是源自情感方法的一些重要概念。

文化社会型实践活动、习惯和传统并不仅仅与现实世界有着一定的联系，而是深深植根于当地的个性与记忆中。在这一层面上，建筑可以作为一种用来与一处空间或是地区情感联系的有效手段。

这些所选工程与纪念物的恢复与世界遗产史不存在任何的关系，但他们关注与过往时光和记忆相关的一些具体遗迹的保护以及向我们日常生活参照所产生的转变。它们之间彼此的相关性来自于作为其历史、记忆和个性的本地或个人文化。

地理学家段义孚曾提到，"时间和空间的相关性是一个错综复杂的问题"，此外，他还提出了几项探究这一主题的不同方法。其中有两个方法被用作我们的此次分析："作为时间的一项功能，某处的附加值就蕴含在这样的一个短语中，即，'对一个地方的了解需要花些时间'，场地成为时间，或成为逝去时间的记忆，这一切变得可能"。（引自《经验透视中的空间与地方》，1977年）

除了这些项目所在地区鲜明的本地特色之外，我们也可以观察到更广阔影响力之下的一些特别之处。本章的项目中有三个位于欧洲（英国、瑞典和德国），一个在日本，这充分显示出在记忆感知上存在的众所周知的文化差异。欧式对记忆的处理手法与实际的物质性更加具有相关性，在这里，设计理念以一种具体的方式加诸一个表皮的建造之上。如此说来，记忆并非那么真实或是物质化，它通过保存理念来实现自身的意义。那是货真价实的体现而并非仅仅是字面意义。

瓦尔特·本雅明在1900年左右所著的一本书《柏林的童年》中曾

Hut in Fichtelberg / AFF Architekten
House C, A Refurbished Hay Barn / Campovono Baumgartner Architekten
Manor House Stables / AR Design Studio
Garden Tree House / Hironaka Ogawa & Associates

Surfaces of Affective Memories / Paula Melâneo

## Surfaces of Affective Memories

Memory is a core issue for history and architecture studies. Our understanding of the past and our awareness of the present made historians and architects to develop theories of what and how should be rehabilitated. For that, strict rules of preservation and conservation were originated, mostly during the 18th and 19th centuries – and some are still effective today.
These ideas contribute to a consistent collective memory of human history at a global level, but are sometimes far from important concepts such as local traditions, sustainability, ecology or, particularly, from affective approaches.
Cultural social practices, habits and customs are not just related with the global reality, but are deeply rooted in local identity and memory. Within this scope, architecture can be a useful device to create affective links with a space or a place.
The selected projects are not related to monuments' rehabilitation or world heritage history, but they concern the preservation of specific remnants connected to past times and memories and their transformation into references of our daily life. Their relevance comes from local or personal culture as significance of its history, memory and identity.

The geographer Yi-Fu Tuan refers *"how time and place are related is an intricate problem"* and suggests different approaches to explore the theme. Two of them are now useful for our analysis: *"attachment to place as a function of time, is captured in the phrase, 'it takes time to know a place'; and place as time made visible, or place as memorial to times past."* (Space and Place: The Perspective of Experience, 1977)
Besides the clear local specificities of the places where the projects are inserted, we can also observe characteristics of a wider cultural influence. Three of the projects are located in Europe (England, Switzerland and Germany) and one in Japan, demonstrating a notorious cultural difference in the perception of memory. The European approach to memory is more related to the physical materiality, where the idea is attached to the construction of an epidermis, in a very concrete way. In the orient, memory is not that physical or tangible, and gets its significance on the conservation of an idea, that is materially represented, but not in a literal way.
Remembering Walter Benjamin's words in his book *Berlin Childhood* around 1900, *"Memory is not an instrument for surveying the*

100年历史的草仓的木质隔墙拉回了过去的记忆
100-year-old barn wooden partition recalls the memories of the past.

提到,"记忆是用来传递其戏剧化效果而非其过往的一个工具"。我们可以把这些项目当作建筑是如何激活一个全新阶段的范例。

在这个项目中,一座现有的马厩建筑,庄园马厩,曾是用来安置赛马的小屋,处于即将垮塌的状态,对于AR设计工作室的建筑师来说,建造一座有着3间卧室的房屋却是一个难能可贵的启发。如果一方面对于材料的重复使用有一个切实且可持续的方法——以及随之而来的能量储备方法——那么在另一方面便存在一个保留施工记忆,以作为情感经历的明确意愿。在总体设计中,老旧的马厩建筑可以被设定出一个方案,并且还是英国南部乡村的一幅动人画卷。这给一些有名的英国作家(诸如查尔斯·狄更斯)带来了极大的灵感。对于内部空间的布局,在最初的布局中就有一个创造性的解决办法。为了呈现出这里之前的用途,一些原有的板材和金属隔墙被保留下来,老旧的木门也以一种极其浪漫的方式再次投入了使用,而马槽变为洗脸池,牵引环改造成为毛巾架。就这样,这栋全新的房屋保留了鲜活的马厩记忆,同时在建造细节上也体现出传统工艺的技法。

当下,在全球化理念和文化中,传统正在渐渐隐退。这使得本地的记忆片段的保留变得更加必要,正如瑞士山脉南部的上瓦莱州的范例。瑞典的Camponovo Baumgartner建筑事务所别出心裁地把一个有着一百年历史的草仓改造成一个被他们称为C号房的家庭周末度假别墅。之前这个草仓用作一间空置的储备室,可以容纳一座住宅。建筑师没有领略到瑞典本地的传统或是乡村建筑,而是使用了一种现代的语言并且尽心地进行设计,来形成一种鲜明的对照。这种处理方法对于维护这座历史建筑以及使其适应新的规则且满足现今生活方式需求的需要来说,无疑是提交了一份令人满意的答卷。

用作参照的早前空间布局中有这样一些元素,比如中央墙体,它把重点空间进行了分割,也隔离出独立的地下室。原先的暗色落叶松立面继续保留,来作为外表皮,而新建房屋全部使用的是未经处理过的浅色木料。在部分区域新住宅是与立面相连的,而在其他不相连处则形成一处空间,来作为门廊。所有的这些特点无一不使人想起这个草仓的前身作为动物之家和干草存放处的功能,而如今,它已然成为一个舒适的小家。

位于德国东部的、由AFF建筑事务所负责的菲希特尔贝尔格的木屋是在一幢1971年建造的木质度假房上进行改建的项目。之后,那里还被用作一个滑雪运动俱乐部的更衣室。20世纪90年代,这里还一度遭到了弃用,而一座全新的小屋正是在这样一处破败不堪的地方诞生的。之前结构的墙体作为新混凝土建筑物的框架,老旧的木质墙体

*past but its theater*", we can consider these projects as examples of how architecture can activate a new stage for experimentation. In the project, Manor House Stables, an existing stable block – old lodge for racing horses – in a run-down state, represented a fantastic inspiration for AR Design Studio's architects to build a 3 bedroom house. If on one hand there is a pragmatic sustainable approach on the re-use of materials – and consequently energy saving –, on the other hand there is an expressed willingness to maintain the construction memory as an affective experience. In the general design, the old stable block is assumed as a scenario and part of the picturesque image of the southern English countryside, which inspired some of the renowned British writers (such as Charles Dickens). For the interior spaces' organization, a creative solution was found from the original layout. With a strong narrative drawn to remind the former usage by preserving the original wood and metal partitions, re-using the old wooden doors and, in a romanticised way, by transforming the horse troughs into sink basins and the tethering rings into towel holders. This new house keeps the memory of the stables alive at the same time shows the expertise of the old craftsmanship in the construction details. As the tradition is losing place to global ideas and culture, the necessity of maintaining some pieces of vernacular memory grows, as this example in the Upper Valais, in the South Switzerland mountains. The Swiss studio Camponovo Baumgartner Architekten remarkably converted a 100 year-old barn into a family weekend house, which they call House C. The barn was used as an empty container where the house fits inside. Without getting into Swiss vernacular or rural architecture, they made use of a contemporary language and depurated design as an evident contrast. The solution provided a successful response to the need of maintaining this historical structure and adapting it to new regulations and demands of today's lifestyle.

Some elements are from the original spatial organization where considered as references, like the central wall dividing the main spaces and the separated basement. The original dark larch facade was kept as an outer skin, where the new house is inserted, entirely constructed with untreated light color wood. In some sections this new house is attached to the facade and in others detached, creating a space between, as a loggia. These characteristics recall the former functions of the barn as a shelter for animals and hay, and now the shelter for a comfortable home.

The Hut in Fichtelberg, situated in eastern Germany, a project by AFF Architekten, is constructed over an original wooden weekend house of 1971, which was afterwards used as a locker room for a sport club in a skiing area. Abandoned during the 1990's, its decay gave origin to a new hut. The walls of the former structure were used as a formwork for the new concrete building. The geometry and texture of the old wooden walls, along with existing doors

照片提供：©AFF Architekten|Sven Fröhlich

之前房屋的混凝土墙体印记使访客产生了家的感觉

Concrete wall imprints from the former house make visitors feel like at home.

的外形和纹理，连同门窗一起为新建的室内混凝土墙体进行定型。就像是指纹一般，老房子的痕迹给新屋的表面也留下了深深的印记，它存在过的唯一记忆就是其身影的消失。通过对"回到根源和基础"这一方法的运用，作为一个"简单和质朴"的基础设施，小屋采取了拥有最小舒适感的设计方案。本地产的云杉木用作地板，而可循环使用的构件和家具对室内进行了完善。"从技术性、功能性以及美学的观点来看，这座小屋低调且稀有，在毫无技术特色的状态下高效地实现了其自身的功能"，AFF建筑事务所的Martin Fröhlich如此解释道。菲希特尔贝尔格的山中小屋所提供给人们的最大的奢侈便是在室内透过朝向森林的大窗来尽情纵览周边。静静地坐在屋内，就在这如水的时光里，我们会默默深思对自己来说生命中的放纵、真实所需和最本真的想法和问题。

花园树屋作为一项设计理念来说是一个极具诗意的记忆处理手法。它是在一幢35年房龄的旧屋边加建的房子，屋内有两棵不能被彻底移除的树。它们是这里以及这个家庭的一段历史。对于Hironaka Ogawa&Associates建筑事务所来说，这两棵树的历史就是这里的"根"，因此它们应该在新建筑中得以保留。无论如何，由于我们对一些重要的元素有一定的记忆和感情，那么这个地方的历史就是属于我们的。这些树在一个特定时期内保留有自身独特的身份——自这座房屋被建以来——一些可以被认为是stabilitas loci（居所之固定）的内容。就像设计师所说的那样，"这些树足足守候了这个家庭35年"。因此，在考虑保留这些树的同时，记忆成为本次设计的主要因素。"要不是出于对这些树的爱和依恋，没有人能够做到这一步"，Hironaka Ogawa解释说。之后，经过脱皮和干燥，它们被放置在先前的位置，但是现如今这里则是位于主结构体之外的增建房屋内部。尽管现在它们已不能再向家人提供阴凉了，枝头上的树叶随风摆动时也不会发出阵阵声响，树木周身所散发出来的气息和色彩也与以往完全不一致。它们仍默默讲述着自己经历过的往事，而未来必然也会见证更多全新故事的上演。

在这些项目中，丰富的记忆和对岁月的感知作为"最伟大的雕刻家"（如Marguerite Yourcenar所说）赋予了新建筑一定的个性和相当特殊的意义。它们让新建筑物安然地矗立在那里，与周围的"根"和谐地融为一体。

and windows, gave shape to the new interior concrete walls. The trace of the former house is imprinted on the surface of the new hut as a fingerprint, and the only memory of its existence is its own absence. With an approach "back to the roots and to the basics", this hut is designed to have the minimum of comfort, as a "simple and spartan" infrastructure, where local spruce trees are the material for the floors and recycled elements and furniture outfit the interior. "Unostentatious and sparse from a technical, functional and aesthetic point of view, the mountain hut performs its tasks efficiently without daring technical features", explains Martin Fröhlich from AFF Architekten. In the interior, the great luxury that the Fichtelberg Mountain Hut has to offer consists in the the views over the wonderful surroundings for contemplation, through some generous windows facing the forest. Sit inside, we can wonder in our thoughts and question what can be excessive in life, or really important and essential, in the "liquid times" which we are living in.

The Garden Tree House is a poetic approach to memory as a design concept. In the site designed for the extension of a 35 year-old house, a pair of trees that could not be left in place existed. But they were a piece of the history of the place and of the family. For Hironaka Ogawa & Associates, the history of these existing trees is part of the "essence" of that place, and for that they should be kept in the new construction. If somehow the history of places belongs to us because we have memories and affections linked to significant elements. Those trees carried a specific identity that was preserved during a certain period of time – since the house existed – something that can be understood as the stabilitas loci (the stability of place). As the architect refers "these trees looked over the family for 35 years". Thus, memory was a main concern in the design, when deciding to maintain the trees. "Nobody would go that far without a love and attachment to these trees", explained Hironaka Ogawa. Then, after being stripped off their bark and dried, they were placed in the same spot, but inside the new house extension as part of the main structure. Even they don't provide a shadow, and sound of their leaves flapping with the wind or have the same smell and color anymore, they have plenty of old stories attached to them and surely will acquire new ones in the future.

In these projects, the richness of memories and the consciousness of time as "the mighty sculptor" (as Marguerite Yourcenar says) gave character and special meanings to the new constructions, pacifying its existence and providing harmony with the place "essence". Paula Melâneo

# 菲希特尔贝尔格的木屋
AFF Architekten

自20世纪早些时期以来，大城市一直都被看作是现代化的实验室。自此，面包切片机、咖啡壶、微波炉、3D对流式烤箱、电磁厨具、水床、电动牙刷、TFT显示屏以及许多其他的技术手段都在这座小屋中找到了自己的位置。使用着地区供暖系统、冷凝式供热系统以及电蓄热炉，建筑可以在每个角落和缝隙都设置理想的温度。普通公寓一词正在渐渐地从建筑师的语言中消失，取而代之的是，人们住在套间、单间公寓、阁楼、顶层豪华套房和排房之类的住宅中。尤为具有奢侈性的一面是，建筑师为实现自我内心需求、满足自身最奇特欲望来改造房屋内部结构的自由度，无论是在中欧的房屋中加设热带沙滩，亦或是在沙漠中滑雪。建筑师利用自己的木屋来容纳数量一直在增加的物体，有些对于他们来说是有用的，有些则是作为纪念品，因为它们的设计极具声望——现代人可都是些慢性子的收藏者。就像德国的人口在减少，但对住房增加的呼声越发激烈这一现状所证实的那样，建筑师需要的是更多的空间。这并非是一个有价值的推断，但这确实提出了一个问题，即建筑师与原始元素及其价值之间的关系究竟是怎样的呢？

本项目的木屋似乎是从周边的景致中截取出的一样，如同一个来自Rittersgrün和Oberwiesenthal马路边，穿过Saxony整个Ore Mountains山脉的混凝土雕塑，文明与自然的边界线就是树木的数量超过房屋数量的地区。这里远离城市，没有任何技术支持、手机信号、洗碗机、微波炉、电视或是冷凝式暖气，混凝土结构就这样向面前的森林张开了双臂。就像公交车站只对街道开放一样，总是朝向预期目的地的方向。与造雪机、滑雪板斜坡或是休闲水疗区大相径庭的是，这里所创造出来的空间使自身远离了现代化生活的动感，也有意识地避免了喧嚣。这个能为6~8人提供餐饮设备的山中木屋很简单，也很简朴，其中细节处的品质展现完全来自于一些可触摸的实物。墙壁和天花板都是混凝土制成的；木地板的材料是用当地砍伐的杉木，而诸如开关、电灯、椅子以及洗手盆之类的配件都是可再生材料；炉子是钢制成的；窗户的设置都按照一定的比例。从技术、功能性以及美学的角度来说，木屋低调且稀有，不带有一丝技术性色彩，有效地运行了自身所承载的任务。它的朴实无华可以满足任何一位远足爱好者的需求，这些远足爱好者认为，一座令人印象深刻的山峰、新鲜的空气以及一顿可口的简餐就是一项基本行程中的全部所需，就像格林童话中的幸运的汉斯一样，对他来说，自由远比财产和财富重要得多。

在木屋的内部，比起面向未来，这里更加面向其源头——木屋保留了之前一处绿锈似的印记。建筑师发现和破解了浮雕结构——如同之前这栋木屋形式以及表面的标记，均由于自身的命运而遭到了抛弃。现在看来，这会让人联想到一个猎人在森林中追踪猎物的画面。平房这种户型最初是被当作周末度假屋而修建的，后来被一个叫"Dynamo"的运动俱乐部作为放置衣柜的房间。它打造了一个实体的底层结构，以形成一个全新的、具有识别性的外部结构，这对于历史的清晰度来说也算是一种参考。对过往的保留显示出这座木屋与底层结构之间的关系——就像一个工艺化的幻灯片一样陪人们从过去走向未来。

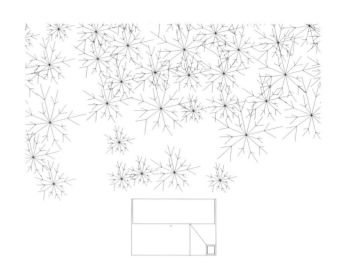

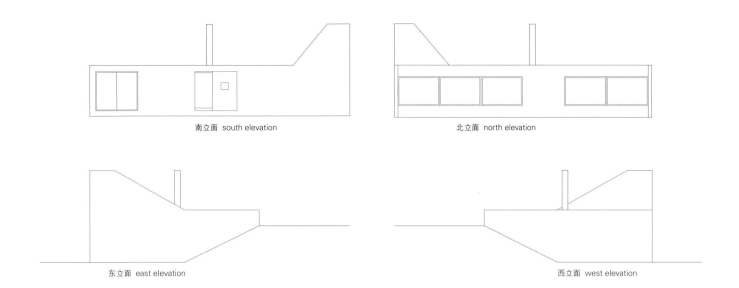

南立面 south elevation　　　　北立面 north elevation

东立面 east elevation　　　　西立面 west elevation

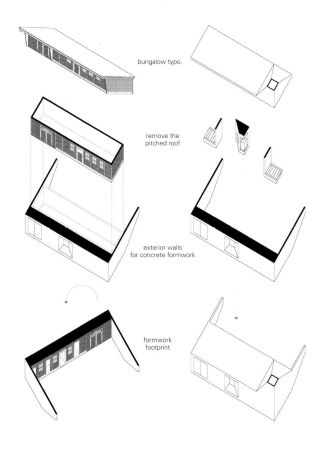

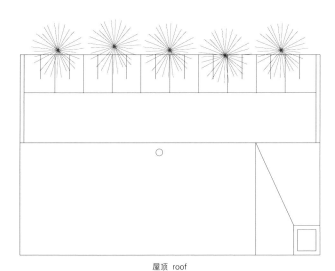

1 房间 2 厨房 3 浴室 4 储藏室 5 露台
1. room 2. kitchen 3. bath 4. storage 5. patio

一层 first floor

由于天气情况比较恶劣,自然状况是菲希特尔贝尔格山脉的一个棘手问题。木屋的功能、使用以及材料都必须按经受住这些考验的条件来进行设计。建筑师在"Hutznhaisl"木屋,这个不会被指控为过气风尚的争论中找到了一个解决问题的方法,在这个动态的系统中,其轮廓并非过度强调其传统性,也不极力显得十分前卫。尽管"Hutznhaisl"木屋代表着与最初当地传统的一种背离,它依然保留着对一个传统木屋类型的一次极具勇气的再次解读。在任何一种情形下,它都对高质量建筑有着诉求,而所有这些,也正是吸引城市人来到这样一座雕刻木质圣诞节装饰比居民数量更多的偏远小山村的魅力之处。

## Hut in Fichtelberg

Since the early 20th Century, the metropolis has been considered the laboratory of modernity. From there, bread slicers, bean-to-cup coffee makers, microwaves, 3D convection ovens, induction cookers, water beds, electric toothbrushes, entertainment on TFT displays and many other technical aids found their way into the huts. Using district heating systems, condensing heating systems and electric storage stoves, the architects set the ideal temperature in every nook and cranny. The common flat is vanishing from their language. Instead, people live in apartments, studios, lofts, penthouses and townhouses. One form of luxury is the architects' liberty to shape the interiors to match their egos, fulfill their strangest desires, whether they involve building tropical islands in Central Europe or skiing in the desert. The architects use their huts to hoard an increasing number of objects, some useful, and some as mementos because of their prestigious design – modern man is a chronic gatherer. The architects need more and more space, as proven by the increase in the demand for housing by a decreasing population in Germany. This is is not a value judgment but it does raise the question as to the architects' relationship with the primal elements and their value.

The hut appears to have been wrestled from the surrounding landscape, a concrete sculpture at the side of the road from Rittersgrün and Oberwiesenthal through the Ore Mountains in Saxony. The border between civilization and nature is where trees begin to outnumber houses. Here, away from the metropolis, there are no technological helpers, mobile phone signals, dishwashers, microwaves, televisions or condensing heaters. The concrete structure opens only to the forest, like a bus stop opens only to the street, always in the direction of the intended destination.

Far away from snow making machines, snowboard slopes or leisure spas, the space created here distances itself from the dynamics of modern life and consciously avoids its busyness. This mountain hut, which offers catering facilities and accommodation for six to eight people, is simple, spartan and the quality of the details is derived solely from tangible things. Walls and ceilings are formed in concrete; the wooden floor boards are made from locally cut spruce trees; fittings like switches, lights, chairs and wash basins are made of recycled components, the stoves are steel and the windows are generously proportioned. Unostentatious and sparse from a technical, functional and aesthetic point of view, the mountain hut performs its tasks efficiently without daring technical features. Its modesty meets the requirements of any hiker, who will confirm that an impressive peak, fresh air and a tasty snack are all they need for an elemental experience, just like Hans in the Grimms' fairytale, to whom freedom was more important than possessions and wealth. Inside, it is more back to the roots than back to the future – the hut preserves the imprint of its predecessor like a patina. The act of decoding and discovering the relief – like markings of the form

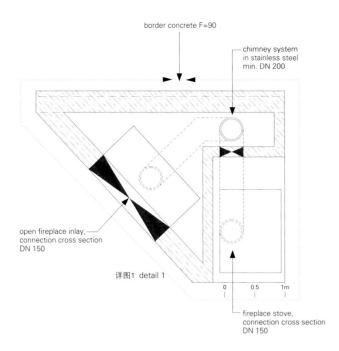

详图1 detail 1

and surface of the former wooden hut, which are abandoned to its own fate, is reminiscent of a hunter following tracks in the forest. The type of bungalow, originally built as a weekend home and subsequently used as locker rooms by the "Dynamo" Sport Club, contributes a mould as a material substrate to form the new, recognizable exterior, a reference to the legibility of history. Saving the past shows the relationship of the hut to its substrate – it accompanies people from the past to the future as a technical lantern slide. With its tough climate, nature is a hard task master on Fichtelberg Mountain. The function, use and material must be designed to withstand these conditions. "Hutznhaisl" seeks a solution within this debate which cannot be accused of being a passing fad. Its silhouette in the dynamic system of its environment is neither emphatically traditional nor does it strive to be avant-garde. Even though "Hutznhaisl" represents a departure from the original local tradition, it remains a courageous re-interpretation of the typology of a traditional hut. In any case, it has the appeal of quality architecture, which has the power to attract townspeople to a remote village with more carved wooden Christmas decorations than residents. AFF Architekten

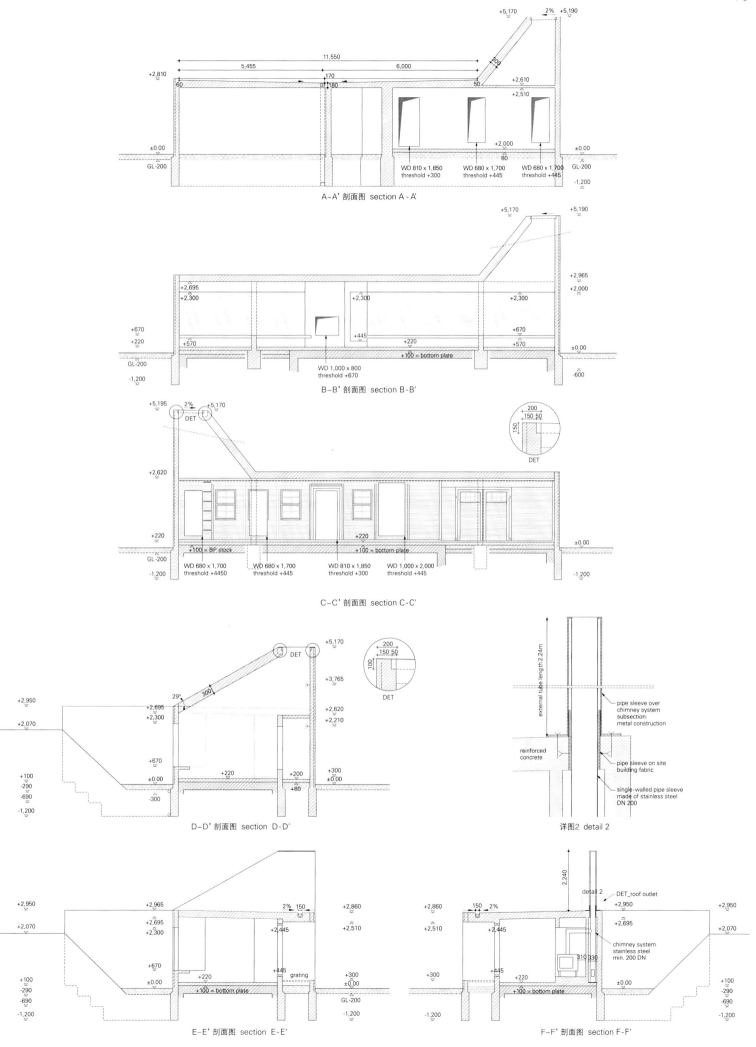

项目名称：Hut in Fichtelberg Mountain
地点：Tellerhäuser Str.5, Ortseingang Tellerhäuser, 09484 Oberwiesenthal
建筑师：Martin Fröhlich, Sven Fröhlich
首席建筑师：Sven Fröhlich
项目团队：Ulrike Dix, Thomas Weisheit
结构工程师：Ingenieurbüro Bauart, Peter Klaus
施工管理：Sven Fröhlich
单价：EUR 729/m²
总楼面面积：82,3m²
竣工时间：2009.12
摄影师：
Courtesy of the architect - p.8, p.9, p.10, p.12$^{top}$, p.13, p.17
©Hans-Christian Schink - p.12$^{bottom}$, p.14

# C号房
一处修缮后的干草仓
Campovono Baumgartner Architekten

百岁的"C号房"坐落于朝向教堂的Reckingen中心地带。Reckingen是这个山谷中的典型居民区。

在这里，住宅和干草仓被分割成不同的单元。较小的干草仓形成紧凑且中间有窄道的深色木质建筑物。这里还建有少量的石屋，有一座巴洛克风格的灰泥立面教堂穿插于其中。与这些木质干草仓相比，身边的教堂就显得极为巨大了。

这座干草仓总共由两部分组成。从结构上来说，这两个单元各自都使用了传统干草仓的布局设计，地面较低是因为之前这里是关放动物的场所，另一处之前被用作干草阁楼的地方则地势较高。

依照最新的动物保护法，这里的所有者必须永久地关闭这个干草仓。将其拆除之后新建一座建筑的想法并不可行，因为这里毕竟是一个历史悠久的村庄中心的一部分，同时也是一个被保护的历史遗迹。因此，当下的任务就是在不破坏其外表的前提下来对其进行改造加工。

干草仓的主框架被保留下来。中间的墙体结构与整个构思同等重要。

建筑师在干草阁楼内构思了一座建筑物，作为周末度假房。内墙立面与其外壳的间距被放大了一倍。就这样，建筑师新建了两间非供暖的凹室，每个干草阁楼里都设有一个，这样就能很好地将其原有的高度以及该空间的木质结构凸显出来。

其中的一个凹室被当作一处入口空间，另一个则放置在起居室前，可以作为一个门廊来使用。内墙表面全部安装玻璃，与外墙相比，无疑是在视角上增大了起居室的空间，并且在新旧元素之间也建立起一种相关性。

开放式楼梯早先是用来将上层的干草运送到下面的干草仓而建造的，现在则成为这座房子的进门楼梯。

度假区包括一处起伏的起居空间和一些个人居住的小房间。房屋中央那片极具历史气息的墙体已经是第二次被拆除了。一个大小、高度以及方向不一的螺旋式通道蜿蜒向上，穿过几间房屋，将两层楼贯通起来。这片区域是纵览群山美景的最佳场所。卧室是相连的，成为生活区的壁龛。最后，所有的门和橱柜都与墙面融为一体。

整个房屋结构都是木质的。地板材质为原生落叶松，墙体和天花板都嵌有高档桦木，这与"木屋"立面较深的旧木材形成了鲜明的反差。屋顶上刚刚落满了落叶松，所有表面都没有经过丝毫的处理，完全任由其自然腐化。

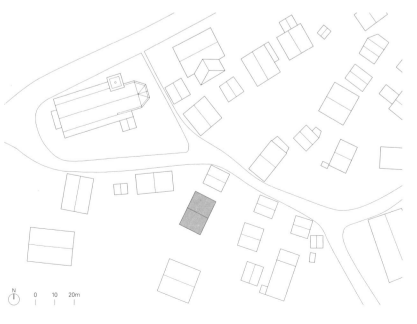

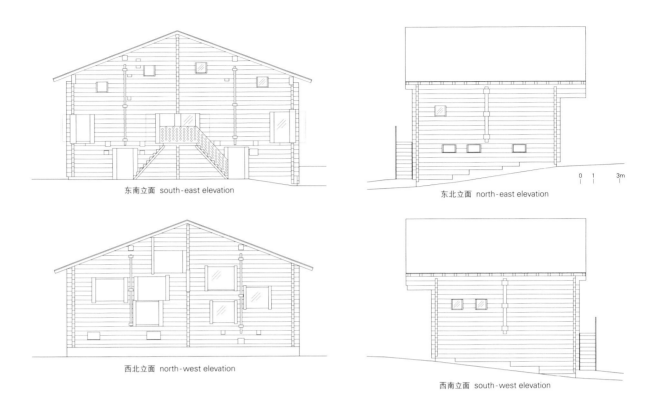

东南立面 south-east elevation   东北立面 north-east elevation

西北立面 north-west elevation   西南立面 south-west elevation

二层 second floor

一层 first floor

| | | | |
|---|---|---|---|
| 1 储藏室 | 6 图书馆 | 1. storage | 6. library |
| 2 建筑设备 | 7 厨房 | 2. building equipment | 7. kitchen |
| 3 入口凹室 | 8 起居室/餐厅 | 3. entrance alcove | 8. living/dining room |
| 4 房间 | 9 起居室凹室 | 4. room | 9. living alcove |
| 5 浴室 | 10 用于卧室的壁龛 | 5. bathroom | 10. sleeping niche |

地下一层 first floor below ground

## House C

The 100-year-old "House C" is located in the center of Reckingen, facing the church.
Reckingen is a typical settlement in this mountain valley.
Houses and barns are divided in separate units. The small-scaled barns form close-standing, dark wooden buildings with narrow alleys in between. There are only few stone buildings, among them, the white plastered Baroque church. The church's, scale is enormous, compared to the wooden barns.
The barn consists of two units. Structurally, each of the units utilizes the traditional layout of the barn, with a lower floor, formerly used to house animals, and an upper floor, formerly used as a hay loft.
Due to new animal protection laws, the owner had to permanently close the barn. Tearing down the barn to build a new construction was not possible because it is part of the historic village center and a protected monument. Therefore, the task was to convert the barn without destroying its outer facade.
The main structure of the barn was kept. The structure of the middle wall was as well important to the concept.
The new weekend house was conceived as a building inside the hay lofts. The inner facade is twice detached from its outer shell. Thus, two non-heated alcoves are developed, one in each hay loft, exposing the original height and wood-structure of the space.
One of them serves as an entrance space, the other is in front of the living room and can be used as a Loggia. The inner facades were generously glazed to visually enlarge the living room to the exterior facades and to create a connection between the old and the new elements.
The open stair, in former times used to bring the hay from above to the barn below, is served today as the entrance stair for the house. The holiday unit consists of an undulating living space and small private rooms. The historic wall, in the middle of the house, is twice removed. A spiral path rises through rooms with different sizes, hights and orientations and connects the two floors. This area offers great views of the mountains. The sleeping rooms are joined as niches to the living space. Finally, the doors and the cupboards are integrated into the walls.
The construction consists entirely of wood. The floor is made of native larch, and the walls and ceilings are made of a high quality birch inlay, which contrasts the dark old wood of the "log house" facade. The roof is newly covered with larch shingles. All the surfaces are untreated and exposed to natural deterioration.

项目名称：Casa C
地点：Reckingen, Wallis, Switzerland
建筑师：Camponovo Baugartner Architekten
用地面积：244m²
占地面积：104m²
可居住的空间面积：114m²
有效表面面积：209m²
总体量：618m³
供暖体量：312m³
竣工时间：2012
摄影师：
©Jose Hevia(courtesy of the architect) - p.22, p.25, p.26
©Jose Hevia - p.19, p.20~21, p.24

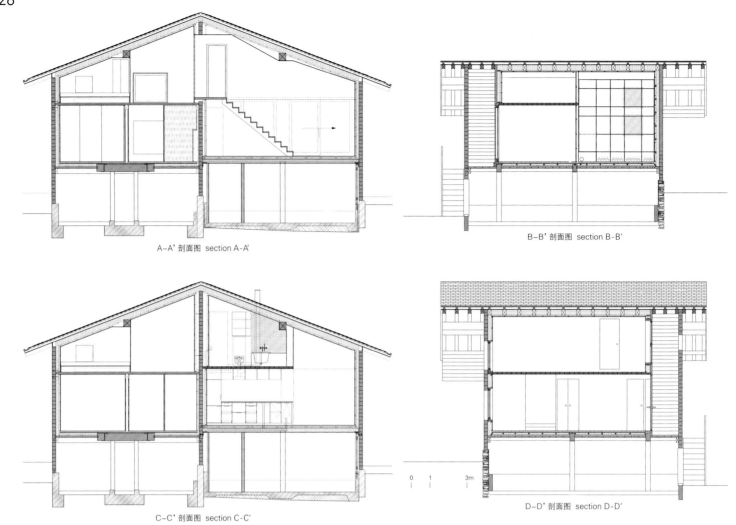

A-A' 剖面图 section A-A'

B-B' 剖面图 section B-B'

C-C' 剖面图 section C-C'

D-D' 剖面图 section D-D'

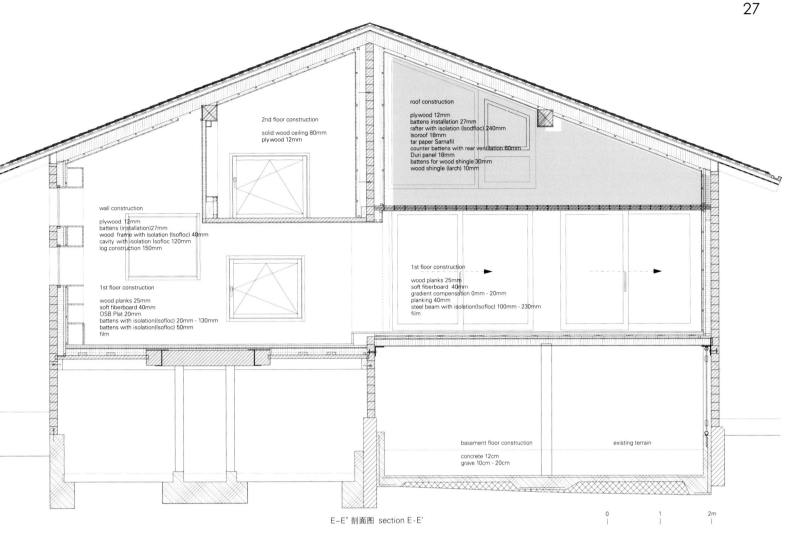

E-E' 剖面图 section E-E'

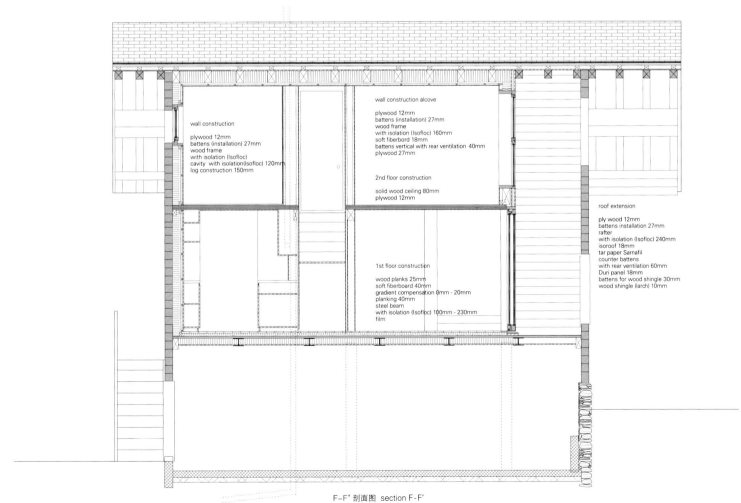

F-F' 剖面图 section F-F'

# 庄园马厩
AR Design Studio

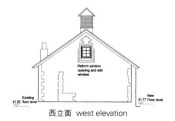

西立面 west elevation

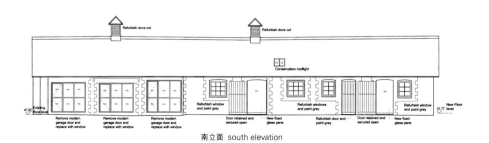

南立面 south elevation

1946年4月5日，一个阳光明媚的周五下午，一众人群在为一场25/1的赛马而欢呼雀跃着。"可爱小草屋"最终欢快地大步跨过了终点线，成为英国最高级别赛马比赛，即越野障碍赛的冠军得主。其训导员是Tommy Rayson，骑手为Robert Petre上尉，这也是它第一次在二战后自1940年以来所参加的真正意义上的安特里越野障碍赛（源自1876年的一个传统比赛，而本次比赛是最后一届在周五举办的赛事）。

就在那个周末，"可爱小草屋"回到了靠近Winchester，名为Headbourne Worthy的小村庄的家中。还没来得及在自己的小窝庄园马厩里站稳脚跟，它就受到了凯旋英雄般的热烈欢迎。

这些一度漂亮且功能强大的马厩已然被闲置了好一段时间，当下甚至有些摇摇欲坠。值得庆幸的是，这个二级列管马厩建筑饱经沧桑的往事并没有被人们所遗忘。RIBA大奖得主，AR设计工作室已经把它完全改造成为一个具有现代优雅时尚气息的三卧之家。

主管Andy Ramus是在庄园马场进行的一次大规模的重建工作时发现了这个被人们所遗忘的历史遗迹。他当下立刻意识到这里具有极大的开发潜力。AR团队完全忽略了其破败不堪的现状。在他们眼里，风景如画的Hampshire乡下地区里的这样一幢拥有丰厚历史底蕴的房屋是完全具有被打造成一幢设计复杂的现代化小屋的潜力的。

马厩的历史和特色成为设计过程中的一个驱动力。AR设计工作室的设计秉持着这样一种理念：设计上的束缚和限制往往会带来最有趣的解决方式。其概念就是在加入简化的创新元素的同时也要将原有的元素妥善保留下来，这样就可以使得原有元素继续熠熠生辉了。这一结果对于马厩的现有布局来说是对其空间的一项创新型布局。如此一来，许多木质的室内墙体才得以展露真颜。接下来，这些墙体经过清洁、剥落和整修，最终显示出其精湛的细节及工艺。

在现有内墙再次焕发风姿之后，下一个任务就是把马厩改头换面，成为时下现代化的家庭式住宅。为了体现出对原有房屋的敬意，设计师对余下设施的改良采取了一种无污染、与时俱进并且较为中和的方式，使其与早先的木质墙体完美地并置一起，它们凸显在极具现代气息的背景内。许多已有的特色之处在这片居家环境内也都得到了翻新和重新规划：早前的马槽被洗刷一新后改造成为洗脸池，牵马环则用做浴室中的毛巾架，而原来的几个门也成功地保留下来，成为具有真实时代感的一个印记。

整个马厩现有三间宽敞的双人卧室（两间为套房）和一间空间较大的家庭浴室。作为眼光长远的一座单层建筑来说，整个设计像是为其量身定做的一般，室内始终贯穿着一条能够将卧房区与起居区完全划分的通道。热情宽敞的开放式厨房及就餐区设置在这一家的中心地带，十分便捷。从这里可以一直通向明亮宽敞的休息室。当然，这还得益于朝向外界宁静村庄的几扇玻璃门。

整座房屋的封闭效果非常好。可加热的抛光混凝土地板对于整体空间来说体现出了功能化的一致性，同时也再次展示了马厩的农用历史。崭新的窗户和顶灯给整个室内营造出一种温馨、明亮和整洁的氛围；形成一处绝佳的环境，来作为一个家庭最好的背景。

已经完工的马厩已然褪去了其破败不堪的旧貌，现如今完全成为一座家住房屋。其每一处都闪耀着当下时代潮流与令人愉悦的年代感这两种气息。

## Manor House Stables

Friday April 5th 1946, on a beautifully clear spring afternoon crowds cheered as the 25/1 racehorse, "Lovely Cottage", strode triumphantly past the finishing post to be crowned winner of the Grand National, the UK's largest horse race. It was trained by Tommy Rayson and ridden by Captain Robert Petre at the first true Aintree Grand National race since 1940, after the Second World War, and the last to take place on a Friday, which had been the tradition since 1876.

That weekend "Lovely Cottage" returned home to the small village of Headbourne Worthy, near Winchester. He received a hero's welcome before settling in for a well-earned rest in the stables at the Manor House where he was housed.

These stables, that were once beautiful and functioning, since remained unused, have fallen into a state of dilapidation. Fortunately, this Grade 2 listed stable block was steeped in poignant historical character and its narrative was not forgotten. It has been transformed into an elegant and contemporary 3 bedroom family home by RIBA award winning architects AR Design Studio.

Practice Director, Andy Ramus, discovered this piece of overlooked historical heritage while undertaking a large scale refurbishment at the Manor House and immediately recognized its potential. The team at AR could see past its existing rundown state. There was a clear potential to create a sophisticated, contemporary family home within the historical context of the building in the picturesque Hampshire countryside.

The history and character of the stables were very much a driving force in design and there is a firm belief at AR Design Studio

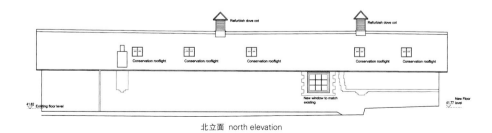

北立面 north elevation

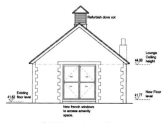

东立面 east elevation

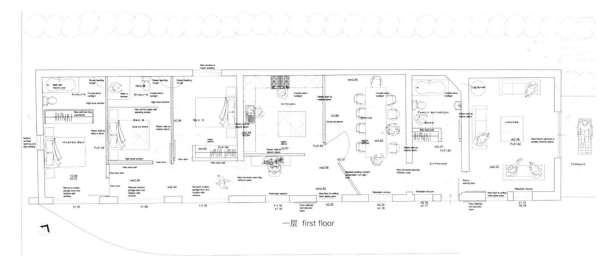

一层 first floor

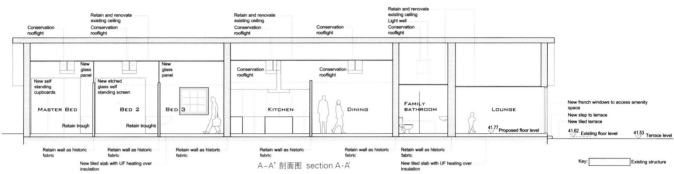

A-A' 剖面图 section A-A'

that design constraints and restrictions can often create the most interesting solutions. The concept was to preserve the existing while making any new additions simple and pure in order to let the original character shine. This results in an innovative arrangement of spaces according to the stables' existing layout, in order to maintain many of the existing exposed timber interior walls. These were then cleaned, stripped back and refurbished to reveal an exquisite amount of detailing and craftsmanship.

With the existing internal walls brought back to life, the next task was to turn the stables into a home for the modern family and bring it into the present day. In order to respect the character of the property a clean, contemporary and neutral approach was taken to the rest of the renovation which juxtaposes perfectly with the original timber walls, allowing them to stand out as pieces of art against a beautifully simple contemporary backdrop. Many of the existing features were refurbished and re-purposed for use in the home environment: the original horse troughs were cleaned and converted for use as sink basins, the old horse ties act as towel rings in the bathrooms and original doors are preserved where possible to give a sense of real period character.

The Stables benefits from 3 large double bedrooms, with 2 suite rooms to accompany a spacious family bathroom. Being a single story property with long continuous views, the layout was tailored and split between sleeping and living accommodation with a single constant circulation running through the entire building. The welcoming and spacious open-plan kitchen & dining area are conveniently located at the heart of the home, leading into the light and roomy lounge which benefits from full height glazed doors that open out onto the sleepy village setting.

The entire property is super insulated, and the heated polished concrete floor throughout provides a functional uniformity to the spaces as well as recounting the stables' agricultural history. New windows and roof lights fitted throughout give the whole place a warm, bright and clean feel, creating an excellent environment as a backdrop for a family home.

The finished stables are completely transformed from its existing dilapidated condition and now a perfectly working family home, bursting with contemporary style juxtaposed against delightful period character.

项目名称：Manor House Stables  地点：Winchester, Hampshire
建筑师：AR Design Studio
项目团队/项目管理：Andy Ramus, Laurent Metrich
结构工程师：Stephen Penfold Associates
室内面积：130m²  室外面积：162m²
承包商：Mike Jacobs Builders Ltd
巡视员/建筑管理：Mike Betteridge/JHAI Limited
竣工时间：2013.3
摄影师：©Martin Gardner(courtesy of the architect)

# 花园树屋
## Hironaka Ogawa & Associates

这是为房主女儿及其丈夫的居住而在一幢有着35年房龄的房屋旁建造的一个扩建项目。

自从35年前建造这幢房屋以来，这里就有一棵榉树和一棵樟脑树。设计要求中提到要把这两棵树移开，因为若不把它们移走，新建的建筑就无法开工了。在建筑师接到这项工程任务但还没到达现场之前，他的脑海中就浮现出了各种版本的设计方案。然而，当他亲眼感受了这一切之后，他所有的想法都立即烟消云散了。

这两棵树就那样浑身充满生命力地站立在他的面前。

建筑师仔细聆听了许多动人的故事。主人的女儿有许多孩童时期攀爬这些树的记忆。这些树陪伴着房主一家一起走过了35个春夏秋冬。它们装点着这座花园，与这一家人共同成长着。鉴于此，继续保留这些树的同时也为客户创造一处新的空间就成为建筑师这次设计的主题。

在细节的处理上，建筑师在确保这些树木枝干毫发无损的前提下将其砍下，之后，他通过连续两周的熏制以及干燥的方式来减少其体内的水分。在这之后，他把它们放置在之前的位置上，并且将其作为起居室、餐厅和厨房中心的主要承重柱子。

为了让它们与之前的成长方式保持一致，建筑师把整个新增建筑体下降了70cm。这样一来，新建筑的高度就低于主房体，但却依然能够保留有4m高的天花板高度。

整个熏制和干燥的过程都在香川县的干燥室内完成。就这样，这两棵树就在自己的家乡重返自己原有的位置。

为了驱除晦气，客户要求在砍伐树木的同时在附近的圣祠里办一场神道教的祭祀。如果不是出于对这两棵树的喜爱和眷恋，相信是没有人会做到这一步的。

在未来，当这座房屋进行拆除，且另一座新建筑被客户的子孙后代在百年后建起来的时候，这两棵树还会以某种形式进行再利用。

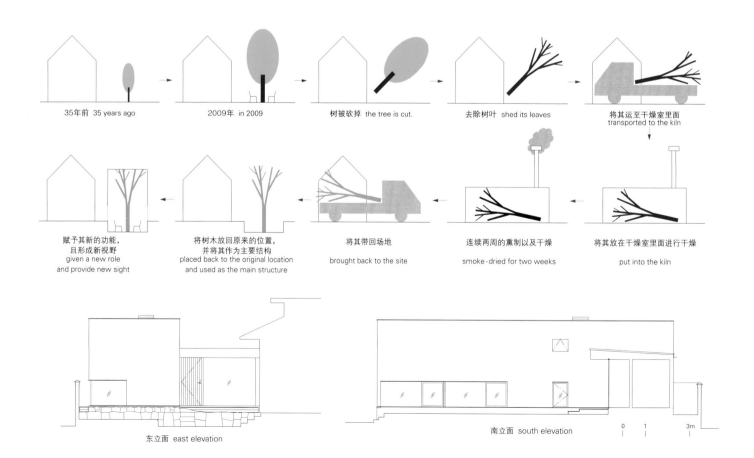

项目名称：Garden Tree House
地点：Kagawa, Japan
建筑师：Hironaka Ogawa
用地面积：699.54m²
建筑面积：50.90m²
有效楼层面积：50.90m²
竣工时间：2010
摄影师：©Daici Ano (courtesy of the architect)

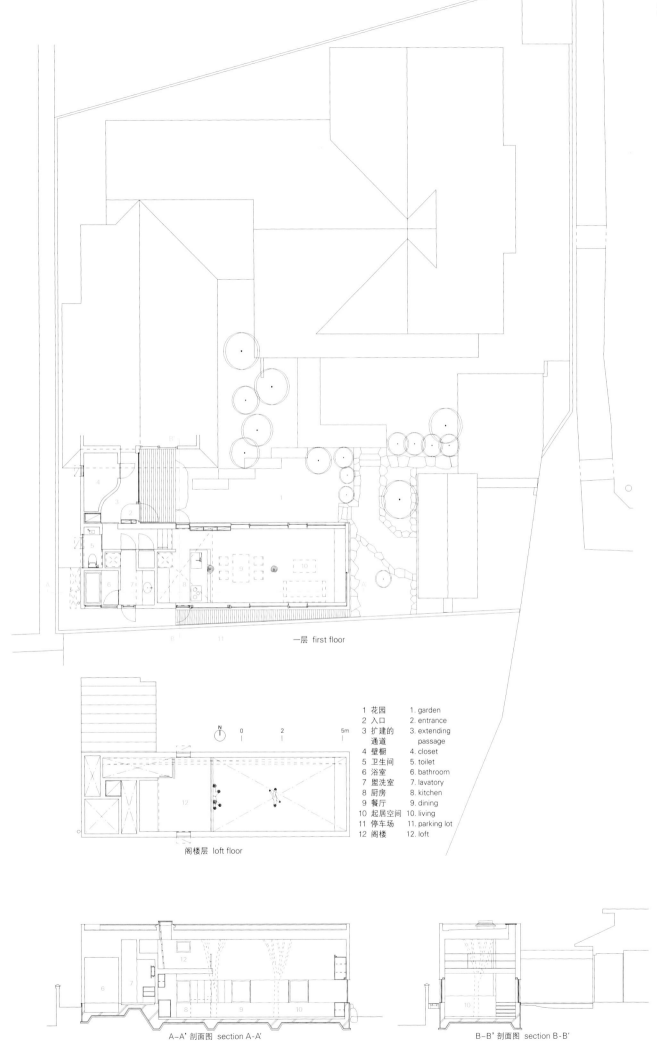

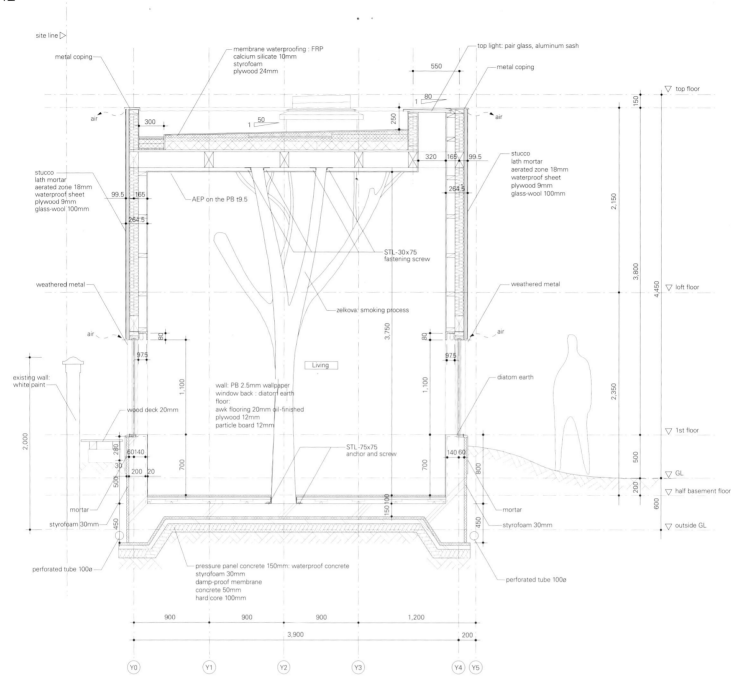

## Garden Tree House

This is an extension project on a thirty-five year-old house for a daughter and her husband.

A Zelkova tree and a Camphor tree stood on the site since the time the main house was built thirty-five years ago. Removing these trees was one of the design requirements because the new additional building could not be built if these trees remained. When the architect received the offer for the project, he thought of various designs before he visited the site for the first time. However, all of the thoughts were blown away as soon as he saw the site in person.

The two trees stood there quite strongly.

The architect listened to the stories in detail; the daughter had memories of climbing these trees when she was little. These trees looked over the family for thirty-five years. They colored the garden and grew up with the family. Therefore, utilizing these trees and creating a new place for the client became the main theme for the design.

In detail, the architect cut the two trees with their branches intact. Then he reduced the water content by smoking and drying them for two weeks. Thereafter, he placed the trees where they used to stand and used them as main structural columns in the center of the living room, dining room, and kitchen.

In order to mimic the way the trees used to stand, the architect sunk the building addition 70 centimeters down in the ground. He kept the height of the addition lower than the main house while still maintaining 4 meter ceiling height.

By the way, the smoking and drying process was done at a kiln within Kagawa Prefecture. These two trees returned to the site without ever leaving the prefecture.

The client asked a Shinto priest at the nearby shrine to remove evil when the trees were cut. Nobody would go that far without love and attachment to these trees.

When this house is demolished and another new building is constructed by a descendant of the client hundreds of years from now, surely these two trees will be reused in some kind of form.

Hironaka Ogawa & Associates

# 新世界
## ——源于旧工厂场地

虽然新建筑仍然是我们的主要建筑焦点，但我们发现，我们的建筑越来越多地利用废旧的基础设施及被遗弃的工业园区。工业用地最终因各种原因处于废弃状态，同时它们的使用期缩短。我们的工业遗址正不断增加，并呼吁我们采取行动。

本章中所展示的项目是这些工业旧址所能提供的许多发展机遇的掠影。它们也是当前恢复和重建旧工业建筑的详细的社会和政治指标。随着新的建筑地点不断地被开发，人们对于工业用地的保护和重建价值的讨论甚至是争论也在进行着。重建工业用地可能会增进我们的文化认同感抑或是满足当前人们对所谓的"真实性"的渴望。它向我们展示了看待时间和持久性的新视角，并提出了什么样的建筑项目应该放在第一位的问题。可以肯定的是，那些工业用地需要有创意的解读并以新的形式体现。这些场所是已建建筑与可行的新兴想法之间的批判性对话。

Whilst newly built structures remain the primary focus of our building activity, we find ourselves with an ever-growing collection of leftover infrastructures and abandoned industrial complexes out there. Industrial sites end up in a state of disuse due to various events and whilst their expiry periods seem to shorten, and our industrial heritage is multiplying and calling for our reaction.

The projects displayed in this chapter are a snapshot of the many development opportunities these places offer, and clear indicators of the current social and political climate in favour of recovering and reconnecting obsolete industrial buildings. New positions are being developed as we speak and the debate on the preservation and the value of these building sites is opening up. Reused industrial sites could possibly enrich our cultural identity or satisfy our current craving for authenticity. It shows us new perspectives on time and permanence, and raises questions regarding what "programme" buildings should be built for in the first place. What is for sure though is that we are left with places that allow for creative reinterpretations and new forms. Places make a critical dialogue between established and emerging ideas possible.

### Turning Former Industrial Sites into New Realities

Looking at today's building stock, we can notice a definite rise in obsolete industrial facilities and infrastructures. We find an array of wastelands and infill sites to our disposal, with an equal wealth of future development options to recover these spaces or buildings. Their obsolescence is provoked by a range of circumstances: some sites lost their significance in the wake of globalization or the rapid transformation of our industrial production; some due to the transition from socialist to market economies or to the availability of cheap labour alternatives. Other sites were abandoned by industry, railways and ports for reasons of upscaling and in search of continuous infrastructural and technological updates. Also, increased planning and environmental regulations affected areas of commercial activity, and in the most extreme circumstances sites were left behind abruptly, following violence or political instabilities. All of the above mentioned result in many industrial and infrastructural complexes, often only a couple of decades old, remaining obsolete and available for a rethink.

The projects in this chapter illustrate a wide range of positions and show proof that there is currently a great deal of social and political support for projects and investments to reconnect sites that got closed in on themselves or divided up by neighbourhoods. Nowadays, urban designers and architects commonly adopt the position to recover these "terrains vagues" and to put everything to work to overcome the decline of under-utilized places. Although it is still a relatively new development to deal with the ageing industrial remnants surrounding us, it is clear that the approach, which was traditionally applied on historic buildings, is opening up. Whereas former approaches were often restricted to either replacing or restoring, we see more complex and intricate solutions with regards to remodelling and regenerating these disused buildings.

Our industrial heritage represents only a small fraction of the world's "official" cultural heritage; in fact only little over 5 percent on the UNESCO World Heritage List. As this percentage is expected to increase rapidly, industrial heritage is becoming an intrinsic part of the current debate on preservation, together with the different stands that are developing towards it. The debate ranges from preservation with a focus on enriching the city and

丹麦国家海洋博物馆/BIG
瓦尔帕莱索文化公园/HLPS Arquitectos
景观实验室/Cannata & Fernandes Arquitectos
Casa Mediterraneo总部/Manuel Ocaña
Conde Duque建筑/Carlos de Riaño Lozano
Can Ribas工厂的修复/Jaime J. Ferrer Forés

将旧工业用地改建为新世界/Tom Van Malderen

Danish National Maritime Museum/BIG
Valparaíso Cultural Park/HLPS Arquitectos
Landscape Laboratory/Cannata & Fernandes Arquitectos
Casa Mediterraneo Headquarters/Manuel Ocaña
Conde Duque/Carlos de Riaño Lozano
Can Ribas Factory Renovation/Jaime J. Ferrer Forés

Turning Former Industrial Sites into New Realities/Tom Van Malderen

## 将旧工业用地改建为新世界

看看当今的建筑存量，不难发现废旧工厂和废弃的基础设施在明显增加。我们有大量的荒地及腾空地盘。如果我们能重新利用这些废弃的空间和建筑，对于我们未来的发展来说，都是财富。这些废旧工业用地是由不同原因导致的：一些是由于全球化的浪潮使它们失去了存在的意义；一些是由于工业生产的迅速转变；还有一些是因为社会主义向市场经济的转变或是找到了廉价的替代劳动力。另外一些场地被工业、铁路及港口所废弃是因为不断更新的基础设施和技术。同样，新增的规划和环境保护规定也影响到了当地的商业活动。很多有极大危害的工业废址尤为显眼，并随之引发了暴力和政治动荡。上述内容是导致拥有20多年历史的工业废墟和基础设施产生的原因。这些建筑仍处于废弃状态，其利用价值有待于人们进一步思考。

本章所列举的项目大范围地阐述了这一境地，并且证明了当前已有来自社会和政府的诸多支持和投资，来重新连接已经封闭的、或是被街区分割的场地。目前，城市设计师和建筑师们通常会选定一个地点来恢复这些"模糊地带"，并且尽全力来使未完全利用的场地重新充满生机。尽管重新利用对我们周围老化的、残留下来的工厂而言，仍然是一种相对新的发展趋势，但是，很显然，应用在这些老建筑的手法已经打开。然而，以前的方法常局限于替换或修建。事实上，我们可以利用复杂的方法去重塑和重建这些废弃的建筑，并使它们恢复生机。

我们的工业遗产仅是世界"官方"文化遗产的一小部分。事实上，据联合国教科文组织的世界遗产名录数据显示，这一数字仅占百分之五多一点。由于这个百分比预计将迅速增加，工业遗产正在成为一个当下关于遗产保护争论的主要议题。对于这一问题，大家各持不同观点。这一争论的观点有支持着力城市建设，并强调城市可以与历史同在，有更为保守的，以及一些对过度保护的观点敏感或是怀疑的。我们应该始终坚持让旧的东西保持原有的模样吗？旧事物在过去是处于一种连续变化的状态的，那么将它恢复到怎样的状态算是最理想的呢？是保存还是重新利用，这个讨论本身就是一个相对新的议题。一些人认为，这一议题是一个理论障碍，它使我们脱离了之前的基础心态：在前行的路上，使过去成为现在的一部分。

对于许多城市和地区来说，能被列入官方名单是一件获益良多的事情，尤其是谈到旅游业和游客数量的时候。但来自STAR建筑事务

emphasizing that cities can live with their past, to a more conservative and possibly problematic position of over-preservation. Should we proclaim that the old has to remain static? What is the ideal status to restore to when the past has been a continuous change? The discussion itself whether it is good to conserve or re-use is a relatively new one. Some claim that it even brought up a theoretical hinder which is deviating us from our previous default state of mind: to move on and make the past part of the present. For many cities and regions it turned out to be very beneficial when they were added to official listings, especially if it comes to tourism and visitor numbers. But as Beatriz Ramo from STAR warns, it also brings along the risk that "preservation frequently freezes sites at some moment in time that never existed in the first place: it dooms them to be hideous and fake versions of themselves. It romanticizes their original essence and preserves their remains for today. Cities will be giant trophy rooms dedicated solely to the brilliance of our past achievements."[1] Danish National Maritime Museum by BIG is located right next a UNESCO World Heritage Site and one of Denmark's most important buildings, the Kronborg Castle. Given that the new museum could not impinge on the views towards the adjacent castle, it was cleverly unfolded as a submerged building within the container of a disused empty dry dock. The Landscape Laboratory by Cannata & Fernandes Arquitectos, with its new roof and sensitive volumetric reconstructions is also situated within a classified sensitive area of high landscape value, south-west of the historic Portuguese city of Guimarães, Potugal.

Architectural space is inherently related to our cultural identity. Our sense of belonging resides not only in physical elements, nor is culture exclusively founded on a system of objects; it involves subjective processes of identification and a discourse that instills meaning to these elements and objects. Most abandoned industrial sites provoke strong historic and cultural ties and touch the feelings and memories of residents and stakeholders. Colour, culture and memory of these places become determining factors in their recovering. What can be learnt from the role of a certain company or manufacturing process contributed to the social development of a neighbourhood? How can these cultural historic

所的Beatriz Ramo警告道，这同时也会带来了一些风险，"经常性的保护行为会使一些地区的发展有时处于冻结状态，而这在一开始是不会体现出来的：这将注定是可怕的、虚假的。它们使最初的本质浪漫化，而为了今天又维持着剩余的部分。城市就像是巨大的纪念品陈列室，装点着我们过去的辉煌成就。"¹由BIG建筑事务所建造的丹麦国家海洋博物馆正好位于联合国教科文组织的世界遗产之地，以及丹麦最重要的建筑之一，克隆伯格城堡旁边。考虑到新博物馆不能侵犯相邻城堡的视野，它巧妙地以一个废弃的空置干船坞内置的水下建筑形式呈现给世人。由Cannata&Fernandes建筑事务所设计的、有着新屋顶及敏感性强的体量的景观实验室，重建在一处景观价值较高的敏感区域，位于葡萄牙历史古城吉马良斯的西南部。

　　建筑空间在本质上与我们的文化身份相关。我们的归属感不仅仅存在于物理元素中，而文化也不是仅仅存在于一个物体系中；它牵涉到主观的识别过程以及给这些元素或物质赋予意义的话语。大多数废弃工业用地有着强烈的历史和文化联系，它们触及到了居民和利益相关者的感情及记忆。这些地方的颜色、文化和记忆成为修复过程中的决定因素。我们可以从某个公司或是建造过程中对附近地区的社会发展所起的作用中学到些什么？如何在保存这些文化历史属性的同时，利用该地区的再开发潜力？特别是当世界是开放的时候，全球化也成为现实，我们似乎在寻找身份，保留着当地的故事和价值观。这种对真实性的渴望也恰好与一种复兴的"毁灭欲望"理念相符。"废墟，尽管处于一种衰败的状态，但是在某种程度上比我们存在的时间长。从文化的角度来看，我们打开废墟是想让自己从年表支配中解放出来，让我们自己在时间的长河中漂流。废墟是悠久历史中的一个片段，但是废墟是一个有未来的时间片段；它在我们死亡之后依然生存，尽管它也提醒着我们其整体性或完美方面的缺失。"²

　　我们看到被工业和环境灾害影响的地区，还有那些冷战时建造的冗余的基础设施，都受到了越来越多的关注。还有，越来越多的艺术家回归到了毁坏及衰败的主题及形象。现代建筑中的遗迹和废墟在当代影像中极为盛行，这也表明对于逝去的过去，人们有着一种浪漫的迷恋。

　　成功的讲述能使时间不再隐形。在Bob van Reeth的《文化耐久性》这篇文章中，他指出耐用的建筑在本质上是"充满智慧的废墟"。"建筑是为多种可能性而设计的，可以说，为了新的用途，建筑可以在时间的作用下向公众敞开。一个项目中的文化持续性也意味着对文化的尊重，而文化建造了特殊的建筑场地。本章中的第一个实例所考虑的不是建筑本身，而是建筑所处的环境，以及在城市发展和我们的时代中，一个地方的特殊比重。所有建筑都有一个综合价值，因为它与地点和时间相关。耐用的建筑还会考虑到居民及用户的智能和情感需求。由于"视角变化"的原因，建筑真正的文化层面是可以大量地反映

attributes be preserved while making use of the area's redevelopment potential? Especially at times when the world is opening up, and with globalisation increasingly being a fact, we seem to be looking for identity and holding on to local stories and values. This longing for authenticity also coincides with a revived "ruin lust". *"The ruin, despite its state of decay, somehow outlives us. And the cultural gaze that we turn on ruins is a way of loosening ourselves from the grip of punctual chronologies, setting ourselves adrift in time. Ruins are part of the long history of the fragment, but the ruin is a fragment with a future; it will live on after us despite the fact it reminds us too of a lost wholeness or perfection."*²

We see a growing attraction to areas that were affected by industrial and environmental disaster, to the redundant infrastructures of the cold war. Also artists increasingly return to themes and images of destruction and decay. The relics and ruins of modernist architecture prevail in contemporary imagery and indicate a romantic fascination with the absent past.

Good storytelling makes time visible. In his essay *Cultural Durability*, Bob van Reeth points out that durable buildings are in essence "intelligent ruins". *"Buildings designed for a myriad of possibilities will be, as it were, broken open by time in order to serve new tasks. Cultural sustainability in a project also means honouring the culture which has created the particular site. In the first instance, this concerns not the building itself but its context, the specific gravity of a place in urban development, and our own time. All buildings have a collective value because of its relation to a place and a time. Durable architecture is also concerned with the intellectual and emotional needs of inhabitants and users. This truly cultural dimension of architecture is reflected precisely in that 'perspective of change'"*³.

The projects presented here work with these dormant values and the potential of the sites. They reactivate the social connections, and they re-inscribe the buildings through narratives of use and new tasks. Valparaíso Cultural Park by HLPS Arquitectos in Chile turns an old prison, where political prisoners were confined and tortured during the years of dictatorship, into a friendly oasis: "a flowerpot in the Valparaíso hills", as the architects call it. Reversing a place of dread, they opened a once impenetrable site to the city. Certain industrial buildings have a neutral character to them, which allows for a level of flexibility. It makes these buildings, often accidental, cope better with changes. We can recognize this character in the Can Ribas Factory Renovation by Jaime J. Ferrer Forés, where an old textile factory with its rational bay of walls and pilasters was transformed into a civic centre and a series of public spaces. Also the barracks of the Conde Duque project in Madrid with its brick walls and steel skeleton structure allowed Carlos de Riaño Lozano to approach it as a big neutral container to house an

1. Beatriz Ramo, "In the name of the Past: Countering the Preservation Crusades", *Monu Magazine #14 Editing Urbanism*, Rotterdam, Netherlands: Board Publishers, 2011, p.80.
2. Brian Dillon, "A Short History of Decay", *Documents of Contemporary Art: Ruins*, London: Whitechapel Gallery and The MIT Press, 2011, p.11.
3. Bob Van Reeth, "Cultural Durability", *Time-Based Architecture*, Rotterdam, Netherlands: 010 Publishers, 2005, p.114.
4. Herman Hertzberger, "Time-based Buildings", *Time-Based Architecture*, Rotterdam, Netherlands: 010 Publishers, 2005, p.82.
5. John Tuomey, Architecture, Craft and Culture, Edge Series #3, Oysterhaven, Ireland: Gandon Editions, 2008, p.29.

出来的。³这里介绍的建筑项目就有这样的潜在价值及场地潜力。它们重新建立了与社会的联系，通过使用说明以及新的用途，来重新定义建筑。由HLPS建筑事务所建造的智利瓦尔帕莱索文化公园就是把一座古老的监狱，在独裁统治的年代关押和折磨政治犯的地方，变成了一处充满善意的绿洲，建筑师们称之为："瓦尔帕莱索山的花盆"。他们向这个城市开放了一处曾经不可攻陷的恐怖之地。

某些工业建筑的特点是中性的，也就是说具有一定程度的灵活性。这一特点使这些建筑往往在无意间能够更好地应对变化。我们可以清楚地看到Jaime j.Ferrer Forés设计的Can Ribas工厂的修复项目的特点。他把内设合理间距的墙壁和壁柱的旧纺织厂改装成市政中心和一些公共空间。同样，在设有砖墙和钢结构的马德里Conde Duque建筑项目中，Carlos de Riaño Lozano把它变成了一个大型中性容器，来容纳一个宏伟的新项目。在《时效性建筑》一书中，Herman Hertzberger唤起了人们的这样一种意识，建筑功能是暂时的，它甚至可能在落成时就已经失效。因此，他一直在寻找能满足多于一种用途的形式或设计。比起"灵活"这个词，他更喜欢"多价"。在Hertzberger看来，"多价"是指自身形成清晰且永恒的建筑形式，但可以根据不同的诠释而发生改变。他所谈到的形式是有巨大的"能力"的：这种能力指的是一种"潜在可能性"，一种可以给建筑充电的潜力。⁴

关于建筑的永久性或是建筑应该与时俱进的想法也同样被John Tuomey所推崇。在认识到这两种方法有时是必要的同时，他指出，修复是一种压抑的剥离形式，而拆除在另一方面代表死亡。他的工作室不愿做关于保留或是拆除的设计，而是想在这种病态的极端与改变耐久性的目标之间做出平衡。他解释道："旧建筑悬浮在禁锢的过去，新建筑投影在一个动荡的未来，在划分艺术史类别的线性年代表的时候不要考虑时间，我们更乐意接受新老建筑共存于现在的生活。对于今天而言是新的东西，到了明天就不一定是新的了。新建筑在被同化，最终新老建筑彼此甚至在彼此的位置上相互理解。"⁵

Manuel Ocaña设计的Casa Mediterraneo总部吸收了新的现实主义元素。这是一个来自不同的历史时期的激动人心的建筑群，清楚地表现了增建、拆除及改建的过程。这个项目也表明了空置的工业用地是新展示形式和布局汇集在一起的地方。这些新形式通常预示了彻底的突破，在后来的许多博物馆、音乐厅和多功能空间里表现得非常普遍。在把以前的工业用地转变为新建筑期间，建筑和城市设计找到了一个特殊实验室来做实验。这为我们提供了一些地方，让我们对新形式能有创意的重新解读，并有可能在已有想法和新兴观念之间进行批判性对话。这种"潜在可能性"无疑改变了我们的白板理念，而这一方法在几十年前是最被建筑师所青睐的。

ambitious new programme. In Time-Based Architecture, Herman Hertzberger raises the awareness that the programme of a building is such a temporary thing that it might even lose its validity by the time the project is in place. He is therefore always looking for forms or designs suited to more than one application. He prefers the word "polyvalency" over "flexibility". For Hertzberger, the idea of "polyvalency" consists of making forms that are themselves lucid and permanent, but can change in the sense that you can interpret them differently. He talks about forms that speak with great "competence": the competence in the sense of a "potential possibility", a potentiality that charges the architecture's batteries.⁴

The idea of permanence or architecture living in time is also encouraged by John Tuomey. Whilst recognizing that both approaches are necessary on occasion, he points out that restoration can be a depressing form of taxidermy, and that demolition on the other hand stands for death. His studio prefers not to design for preservation or obliteration but to steer between these morbid extremes and aim for a changing kind of endurance. He explains: *"Instead of considering time as dividing in linear chronological art-historical categories, with old buildings suspended in a petrified past and new buildings projected in a volatile future, we prefer to think of all buildings co-existing in the context of the living present. What is new today cannot remain new tomorrow. New buildings become assimilated and are eventually understood in relation to each other and to their place."*⁵.

The Casa Mediterraneo Headquarters by Manuel Ocaña assimilated new realities. It is an exciting collection of buildings from different historical periods, and clear manifestations of a process of addition, subtraction and change. This project also shows the proof that vacant industrial sites are the places where new forms of displays and organisation come together. These new forms are often the precursors for radical breakthroughs which become widespread in many museums, concert halls and multi-functional spaces later on. In turning former industrial sites into new realities, architecture and urban design have found an exceptional laboratory for experimentation. We are left with places that allow for creative reinterpretations of new forms, and make possible a critical dialogue between established and emerging ideas. This "potential possibility" has definitely shifted our approach from the tabula rasa idea, and the approach is most favoured by architects only a couple of decades ago. Tom Van Malderen

# 丹麦国家海洋博物馆
BIG

丹麦国家海洋博物馆必须在独特的历史和空间中定位。它坐落于丹麦最重要也是最著名的建筑物之一与一座全新的、生机勃勃的文化中心之间。在这种环境下建造博物馆，充分证明这一建筑能够理解它在所在地区的角色，特别是在克隆伯格城堡的基础上诠释自己，如同一座在干船坞里修建的地下博物馆。

和60年历史的旧船坞的墙壁毫无接触，整座博物馆位于地下，并且围绕着干船坞的墙壁设置成一个连续的环状——使船坞成为整座博物馆的核心区域——一处能让参观者感受船舶建筑规模的开放式户外区域。

三个双层桥梁跨越整个干船坞，不仅作为都市连接，还为参观者提供了通向博物馆不同区域的捷径。海港大桥在封闭船坞的同时，还能作为"海港大道"；博物馆的礼堂可作为连接附近文化中心及克隆伯格城堡的桥梁；曲折的倾斜桥梁将访客引至博物馆的主入口。这座桥将新观旧貌连接在一起，当参观者向下行至博物馆时，可以俯瞰博物馆地上及地下的壮丽景观。丹麦悠久而荣耀的海洋历史在深达地下7m的船坞内部及周围一一展开。博物馆内所有楼层都将展览空间与礼堂、教室、办公室、咖啡厅及船坞层连接起来——形成坡缓的地势，以创造令人兴奋的雕塑空间。

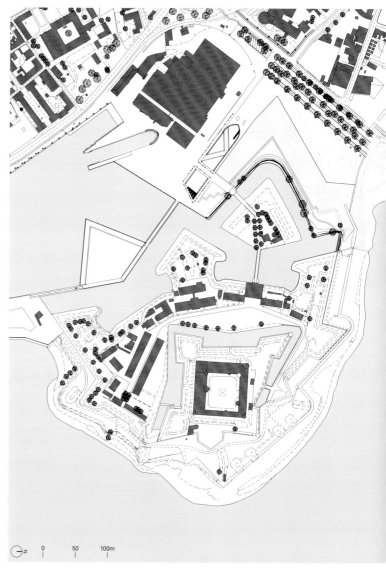

博物馆设置在旧干船坞内,从城堡一处来看,处在联合国教科文组织设置的500m保护线之内
The museum is placed in the old dry dock within the 500m UNESCO preservation line from the castle

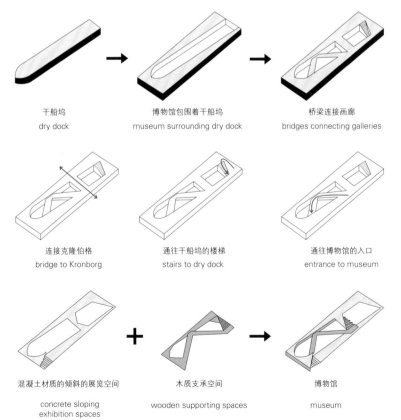

干船坞 → 博物馆包围着干船坞 → 桥梁连接画廊
dry dock — museum surrounding dry dock — bridges connecting galleries

连接克隆伯格 / 通往干船坞的楼梯 / 通往博物馆的入口
bridge to Kronborg — stairs to dry dock — entrance to museum

混凝土材质的倾斜的展览空间 + 木质支承空间 = 博物馆
concrete sloping exhibition spaces + wooden supporting spaces = museum

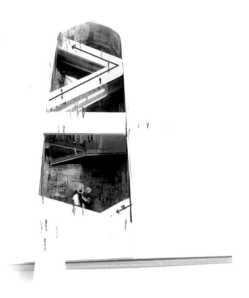

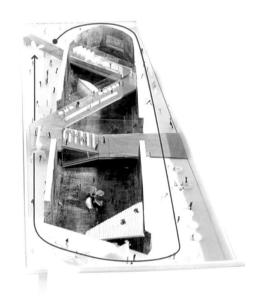

A-A' 剖面图 section A-A'

B-B' 剖面图 section B-B'

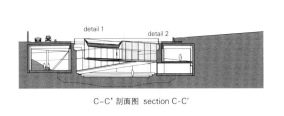

C-C' 剖面图 section C-C'

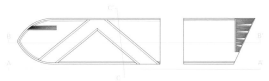

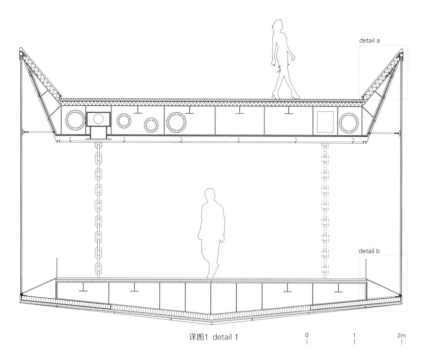

详图1 detail 1

项目名称：Danish National Maritime Museum
地点：Helsingør, Denmark
主要合伙人：Bjarke Ingels, David Zahle
项目主管：David Zahle
项目团队：John Pries Jensen, Henrik Kania, Ariel Joy Norback Wallner, Rasmus Pedersen, Annette Jensen, Dennis Rasmussen, Jan Magasanik, Jeppe Ecklon, Karsten Hammer Hansen, Rasmus Rodam, Rune Hansen, Alina Tamosiunaite, Alysen Hiller, Ana Merino, Andy Yu, Christian Alvarez, Claudio Moretti, Felicia Guldberg, Gül Ertekin, Johan Cool, Jonas Mønster, Kirstine Ragnhild, Malte Kloe, Marc Jay, Maria Mavriku, Masatoshi Oka, Oana Simionescu, Pablo Labra, Peter Rieff, Qianyi Lim, Sara Sosio, Sebastian Latz, Tina Lund Højgaard, Tina Troster, Todd Bennet, Xi Chen, Xing Xiong, Xu Li
甲方：Helsingør Municipality, Helsingør Maritime Museum
用地面积：6,500m²
有效楼层面积(包括船坞的墙体)：7,600m²
设计时间：2007 竣工时间：2013
摄影师：
©Luca Santiago Mora(courtesy of the architect) - p.53, p.54~55, p.56~57, p.59, p.61, p.62
©Rasmus Hjortshoj(courtesy of the architect) - p.48~49, p.52~53, p.63, p.64, p.65 (except as noted)

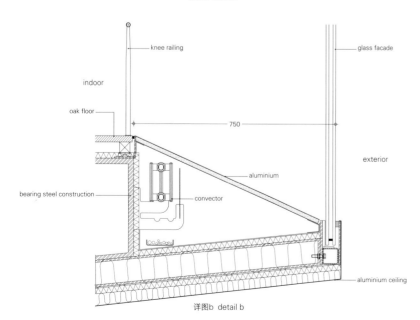

详图a detail a

详图b detail b

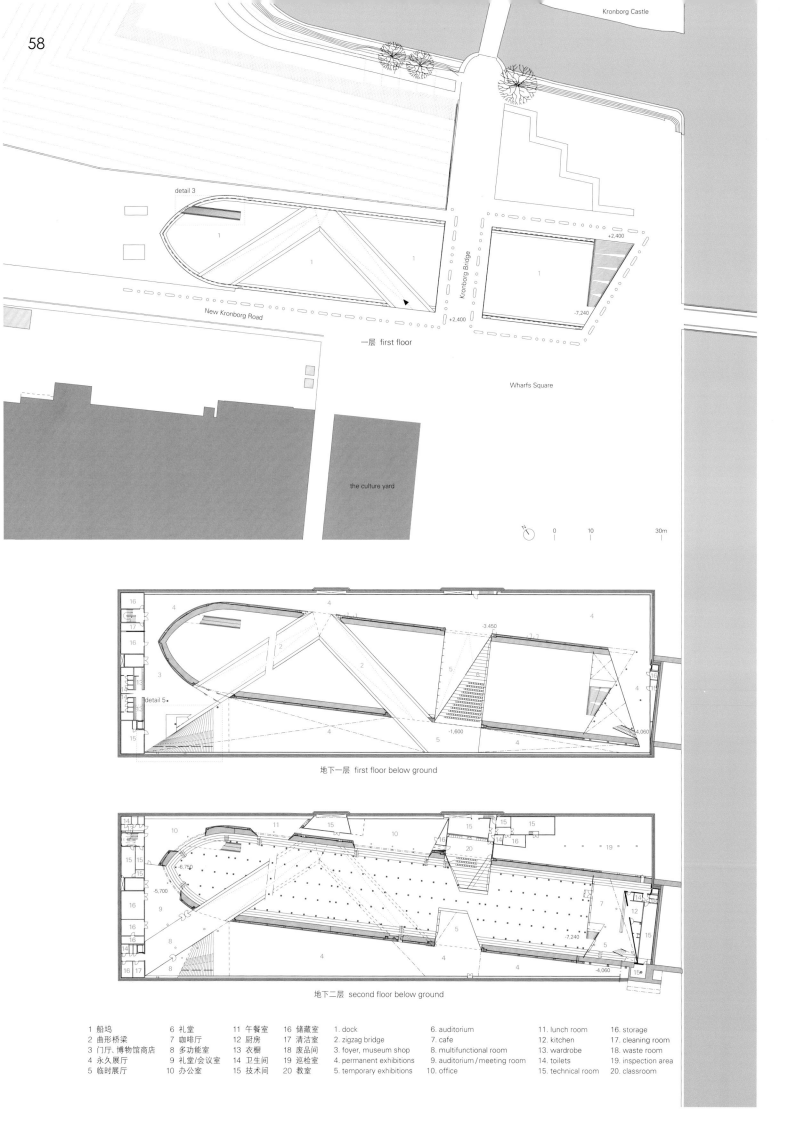

## Danish National Maritime Museum

The Danish National Maritime Museum had to find its place in a unique historic and spatial context, between one of Denmark's most important and famous buildings and a new, ambitious cultural center. This is the context in which the museum has proven itself with an understanding of the character of the region and especially the Kronborg Castle, like a subterranean museum in a dry dock.

Leaving the 60 year old dock walls untouched, the galleries are placed below ground and arranged in a continuous loop around the dry dock walls – making the dock the centerpiece of the exhibition – an open, outdoor area where visitors experience the scale of ship building.

A series of three double-level bridges span the dry dock, serving both as an urban connection, as well as providing visitors with short-cuts to different sections of the museum. The harbor bridge closes off the dock while serving as the harbor promenade; the museum's auditorium serves as a bridge connecting the adjacent Culture Yard with the Kronborg Castle; and the sloping zig-zag bridge navigates visitors to the main entrance. This bridge unites the old and new as the visitors descend into the museum space overlooking the majestic surroundings above and below ground. The long and noble history of the Danish Maritime unfolds in a continuous motion within and around the dock, 7 meters(23 ft.) below the ground. All floors – connecting exhibition spaces with the auditorium, classroom, offices, cafe and the dock floor within the museum – slope gently, creating exciting and sculptural spaces.

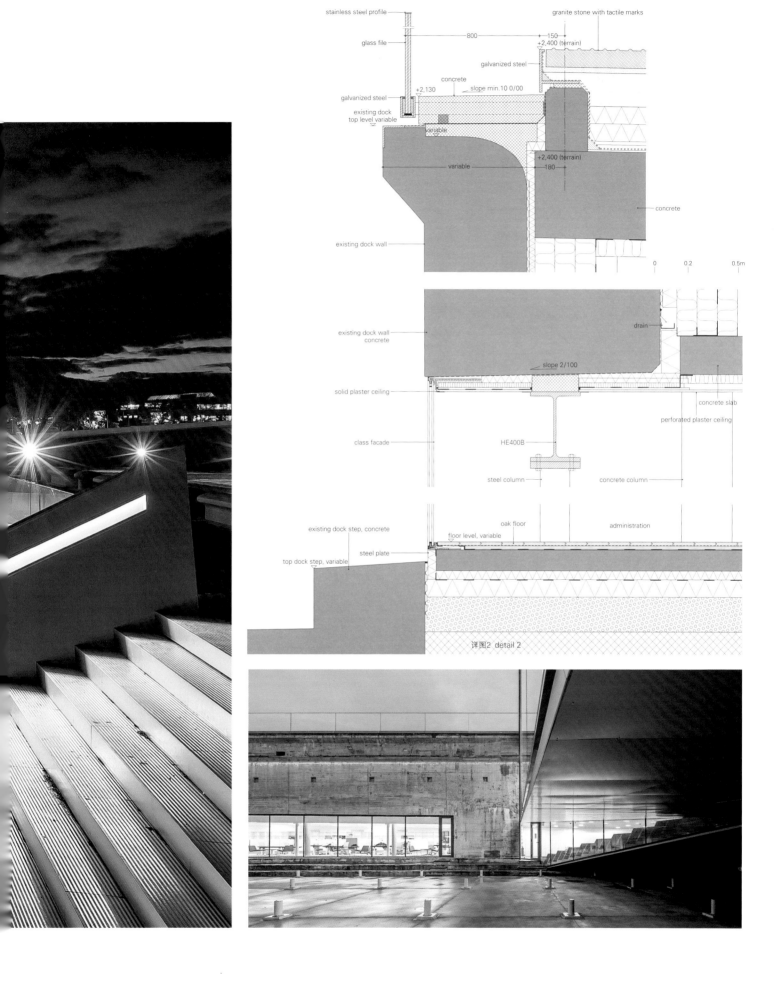

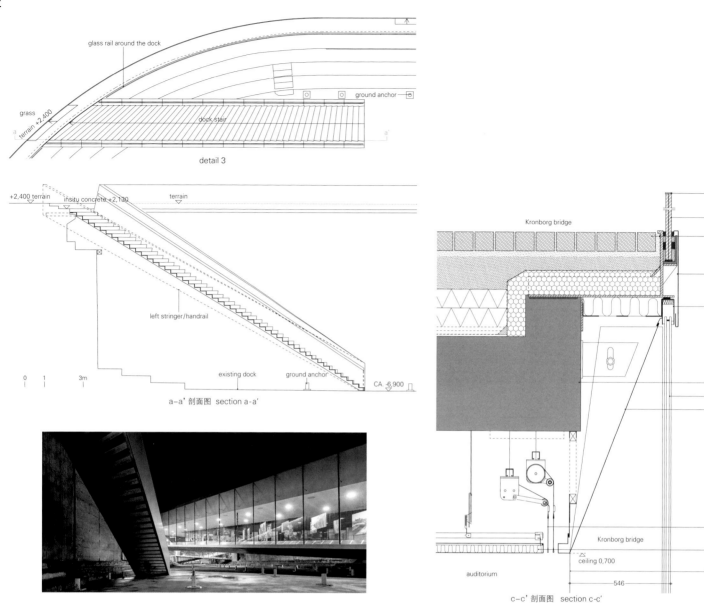

detail 3

section a-a'

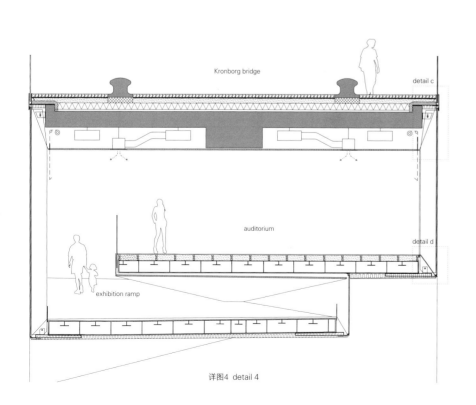

section c-c'

detail 4

section d-d'

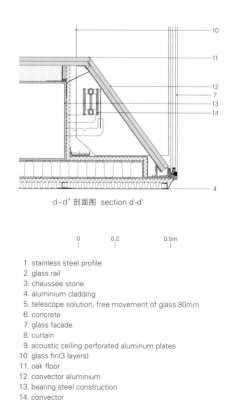

1. stainless steel profile
2. glass rail
3. chaussee stone
4. aluminium cladding
5. telescope solution, free movement of glass 80mm
6. concrete
7. glass facade
8. curtain
9. acoustic ceiling perforated aluminum plates
10. glass fin(3 layers)
11. oak floor
12. convector aluminium
13. bearing steel construction
14. convector

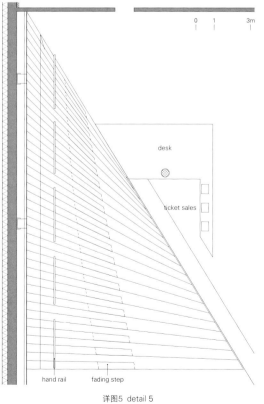

详图5 detail 5

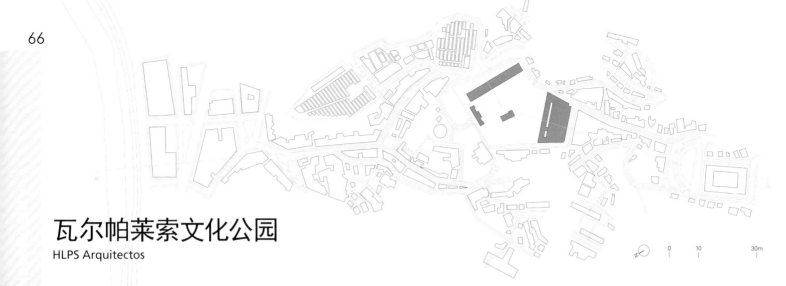

# 瓦尔帕莱索文化公园

HLPS Arquitectos

文化公园位于瓦尔帕莱索山，被看作是整合和集成的新空间。由于其最初的建筑目的是用作监狱，所以它呈现出一种潜在的矛盾：封闭的幽禁空间。建筑的主要问题在于它的单一性。这也可以被看作是它最基本的问题：怎样把密闭的空间转化为整合的空间(最终的建筑作品不在于寻求答案，而在于寻找什么样的问题要解决……把选择转移到提问)；通过识别建筑场所的四个显著且重要的位置——将其转化成一个具体的问题——一系列操作便成为答案，且每个操作都是在不同且相关的地形水平上实施的。

**花盆**

怎样把密集的监狱改建成开放的空间呢？

之前的监狱场地展示为石头平台以及墓地包围的堡垒。鉴于监狱及其作为山上最大水平面的特定条件，建筑师提出了清除这些年来所有轻型建筑结构的建议：只保留监狱长廊建筑、旧入口建筑和西班牙人留下的旧火药桶，把它们放在新场地的显著位置上。清除后的场地设置了常规的网格，种有紫薇花，使场地蒙上了一层面纱；而一群小岛上种有赛波树和玉兰树，以标出一些出众的地点，设有一条草地覆盖的大道，给诸多活动提供了场所；而小棕榈树林则作为主入口的标记；一组古坟内，建筑师将重新利用移走的老建筑的地下室，作为公园的主要规划线。公园仍然由老围墙和走道环绕。

建筑师把建筑原址修建成瓦尔帕莱索山的花盆。

**水平线**

如何把监狱改修成回廊？

+00,0   THE HORIZONTAL PLAN: SURFACE CLEARANCE
**FLOWERPOT**
*How can the dense prison be turned into an open space?*

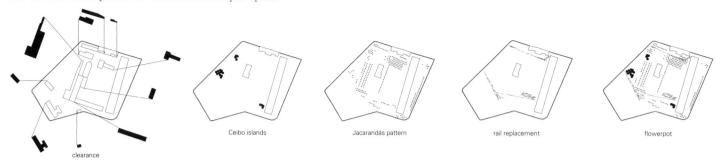

+03.00   CONFINING PERIMETER: THE WALL CUTTING
**HORIZON**
*How to turn the prison confinement into cloistering?*

+04.0   BETWEEN RAVINES: NEW DIRECTION
**PROMENADE**
*How to set a relation between the walled interior and its surroundings which open towards the hills?*

+12.0   GEOGRAPHIC CONDITION: NEW URBAN PLANE
**TRESTLES AND SADDLE**
*How to build a 8.500m² cultural center over a park without taking away public open surface?*

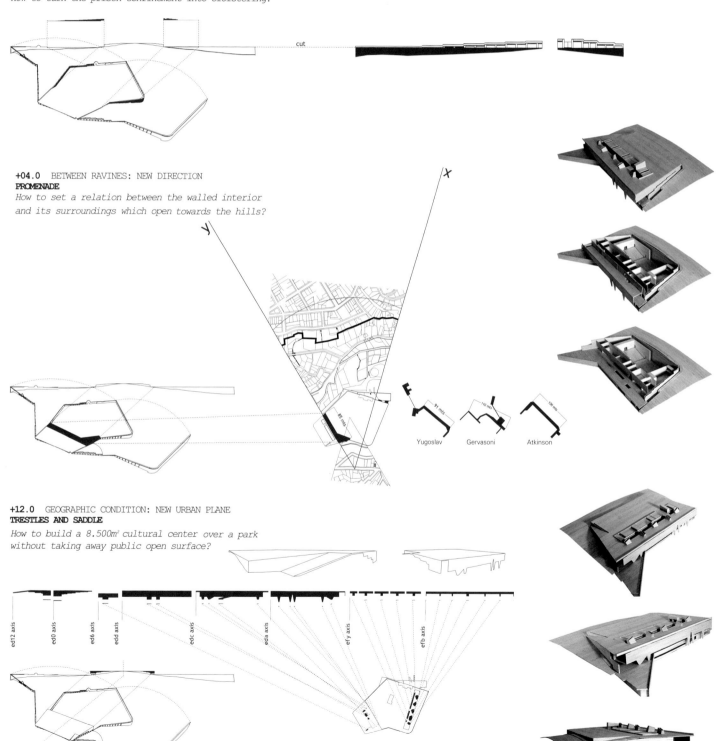

针对现存的重要水平线，建筑师提出了水平切割围墙的方案。

根据相关墙体的水平线，建筑师画出了新水平线。从抽象方面来说，它废除了所有相关的监狱标志和关联性，把监狱改建成回廊。警卫通道、牢房及扶手的所有痕迹都被抹除。墙体变成抽象的结构，仅被当作是一种材料。

这条新水平线，与内部水平线相关联，形成了不同的场景，在封闭的回廊及开放的群山和大海视野之间移动。

**散步道**

如何设置围墙内部及其面对群山的周围环境之间的关系？

新的水平线与卡塞尔山的两个交叉点会出现一种全新的可能性。一条新的人行道连接着峡谷的两边。散步道，也可作为一个阳台，穿过公园，与附近其他的现存通道网络连在一起。

建筑与临近的山丘和历史保护区，如阿雷格里山和康塞普西翁山之间，构建了一种全新的关系。

这个新方向在建筑场地上提供了一个绝佳的方法，第二条地基线与牢房建筑（牢房规模一致）之间建立了一种关联，重新定义了空间的位置，并且在新的中心位置放置古西班牙人的火药桶。

**栈桥和马鞍**

如何在一座面积为8500m²的公园之上建造一个文化中心，而不占用任何公共表面积？

根据尽可能多地释放公共空间表面、保持更多的公园面积及总容积率的理念，建筑项目应从地面抬升起来，并且建在公园之上。为使这些密闭的重型建筑结构与地面以一种温和且有效的方式接触，建筑师建造了一对长方形带孔的结构面，来作为栈桥。这些结构将在高层容纳一些文化功能，并且逐次向下布局公共空间。

为了在这些栈桥上延伸上层场地的水平线，建筑师在峡谷两侧之间修建了一个平面。这个像马鞍一样的结构悬在山腰之上，与低处的公园（向下12m的距离）建立了一种全新的关系，同时也与瓦尔帕莱索港的地理环境和大海形成一种新联系。一个位于文化中心下方的多功能城市化高地便建成了。

这些操作行为，在公共空间的增建及它们的内部水平关系方面，重新定位了围墙，并且为公园、社区及小山规划了一处集成空间。

## Valparaíso Cultural Park

The commission of a cultural park located in the hills of Valparaíso, understood as a new space of confluence and integration, presents an underlying contradiction due to its original purpose as a prison: the hermetic reclusion of space. The main architectural problem lies in this singularity, and can be set as a basic question: how to turn enclosure into an integrating space (composition does not lie on answers but on what questions should be done ... shifting from the responsibility of choosing to the one of asking): Through the recognition of four evident and significant situations found on the site – each turned into a specific question – a series of operations are set as answers, each of them happening on different and relevant topographic levels.

### Flowerpot

How can the dense prison be turned into an open space?

The former prison's site shows as a walled fortress surrounded by stone platforms and cemeteries. Remarking the former prison and its special condition of being the largest horizontal surface over the hills, a general clearing of all the light constructions accumulated along the years is proposed, keeping only the prison gallery building, the old entrance building and the old spaniard's powder keg, putting them in an outstanding situation over this new field. This clearance is planted with a regular grid of jacarandás –

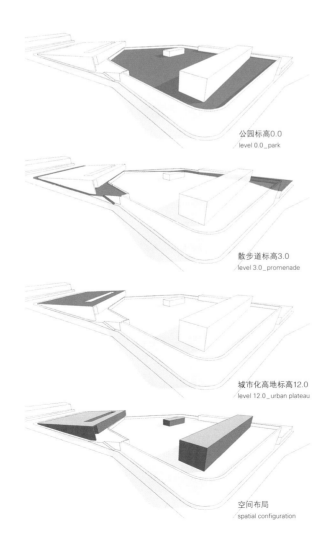

公园标高0.0
level 0.0_park

散步道标高3.0
level 3.0_promenade

城市化高地标高12.0
level 12.0_urban plateau

空间布局
spatial configuration

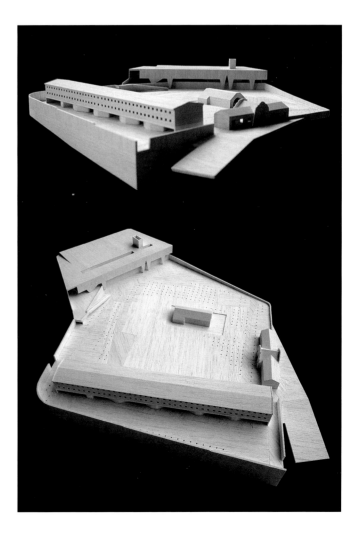

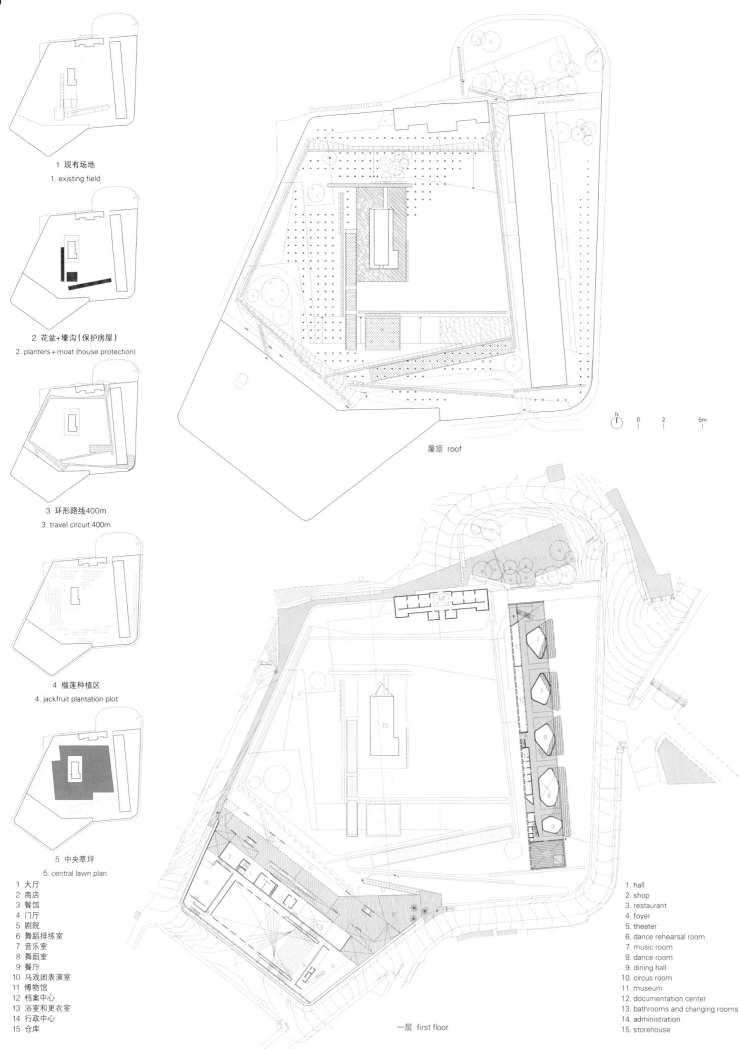

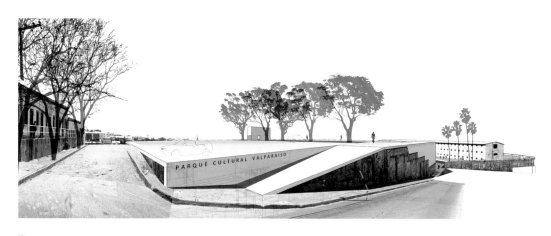

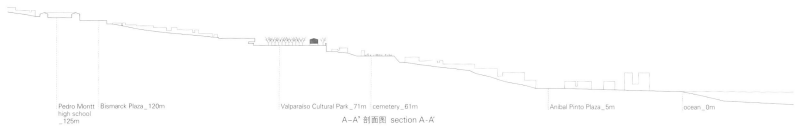

| Pedro Montt high school _125m | Bismarck Plaza _ 120m | Valparaiso Cultural Park _71m | cemetery_61m | Anibal Pinto Plaza _5m | ocean _0m |

A-A' 剖面图  section A-A'

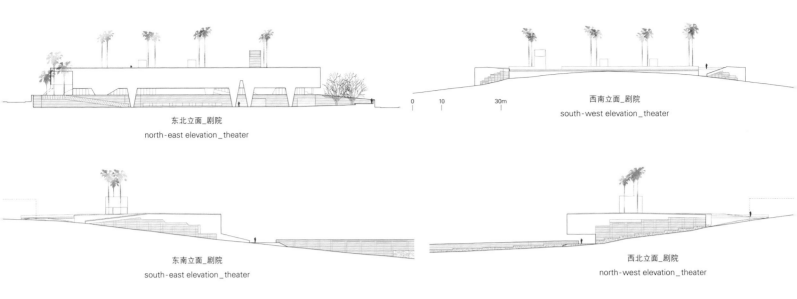

东北立面_剧院
north-east elevation_theater

西南立面_剧院
south-west elevation_theater

东南立面_剧院
south-east elevation_theater

西北立面_剧院
north-west elevation_theater

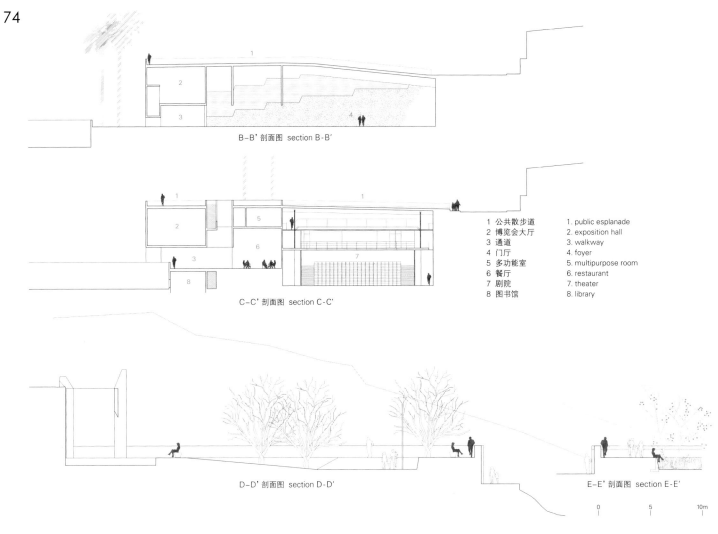

B-B' 剖面图  section B-B'

C-C' 剖面图  section C-C'

| 1 公共散步道 | 1. public esplanade |
| 2 博览会大厅 | 2. exposition hall |
| 3 通道 | 3. walkway |
| 4 门厅 | 4. foyer |
| 5 多功能室 | 5. multipurpose room |
| 6 餐厅 | 6. restaurant |
| 7 剧院 | 7. theater |
| 8 图书馆 | 8. library |

D-D' 剖面图  section D-D'          E-E' 剖面图  section E-E'

项目名称：Valparaíso Cultural Park
地点：Valparaíso, Chile
建筑师：Jonathan Holmes, Martin Labbé, Carolina Portugueis, Osvaldo Spichiger
合作商：Nicolás Frienkel, Jorge Síviero, Carolina Moore
结构工程师：Luis Soler y Asociados
施工单位：Constructora Bravo Izquierdo Ltda.
声学工程师：Carla Badani
剧院音响系统：Constellation. Meyer Sound
照明：Limarí Lighting Design Ltda.
景观建筑师：HLPS Arquitectos, Paulina Courard
用地面积：21,000m² 总建筑面积：8,711m²
材料：concrete, stone pavement
设计时间：2009 竣工时间：2011
摄影师：
Courtesy of the architect-p.71, p.72~73, p.77top, p.78
©Cristobal Palma-p.66~67, p.74~75, p.77bottom, p.79, p.80, p.81

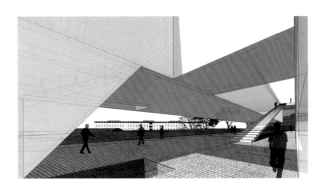

that will build a veil over the site; with a group of islands with ceibo and magnolia trees – that will landmark some outstanding spots; with a grass esplanade – that will house massive activities; with little palm groves – that will mark the main entrances; and with a group of barrows – that will recycle removed old construction's basements as the main park arranging lines. Park remains enclosed by the old perimeter wall along with the walkway that accompanies and embraces it.

The site is turned into a flowerpot on Valparaíso hills.

### Horizon

How to turn the prison confinement into cloistering?

Beginning from a significant existing level, a horizontal cut in the wall that runs around the perimeter is proposed.

From a relevant wall level a new horizon is drawn. In its abstraction it suppresses any reference to prison symbols and references, turning the prison confinement into cloistering. All traces of guard walkways, shelters and hand-railings are removed, turning the wall into an abstract construction, taking it to a barely material problem.

This new horizon – in its relation with inner levels – generates various situations that move between the cloister enclosure and open sights to the hills and the sea.

### Promenade

How to set a relation between the walled interior and its surroundings which open towards the hills?

In the two junctions of the new horizon and the slopes of Cárcel Hill a new possibility appears, a new walkway connecting each side's ravines. A promenade that, as a balcony, crosses over the park and relates to the neighbouring existing walkways network. A new relation with the neighbouring hills and historical conservation zones – like Cerro Alegre and Cerro Concepción – is built.

This new direction establishes a new wise on the site, a second foundational line, in its relation with the prison cells building – with similar dimensions – reorders space and relocates the old spaniards powder keg at the new center.

### Trestles and saddle

How to build a 8,500m² cultural center over a park without taking away public open surface?

According to the idea of liberating as much surface for public space as possible and keeping the park area to its total capacities, built programmes are lifted from the ground and built over the park. Allowing these structures – hermetic and heavy on top – to reach the ground in a mild and efficient way, a couple of oblong and punched structural planes are proposed in a trestle way. These structures will house cultural programmes on the higher levels and will order the public space on their downfall.

Extending the upper site level over those trestles a plane that swings between the sides of the ravine is set, a sort of saddle over the hill loin establishes a new relation with the lower park – 12 meters down – and with Valparaíso's harbour's geographic context and the sea. An urban multipurpose plateau is beneath where the cultural center is built.

These operations, in their public space multiplication and in the relation between their own levels, reorient the enclosure and propose a space of integration for the park, the neighbourhood and the hills.

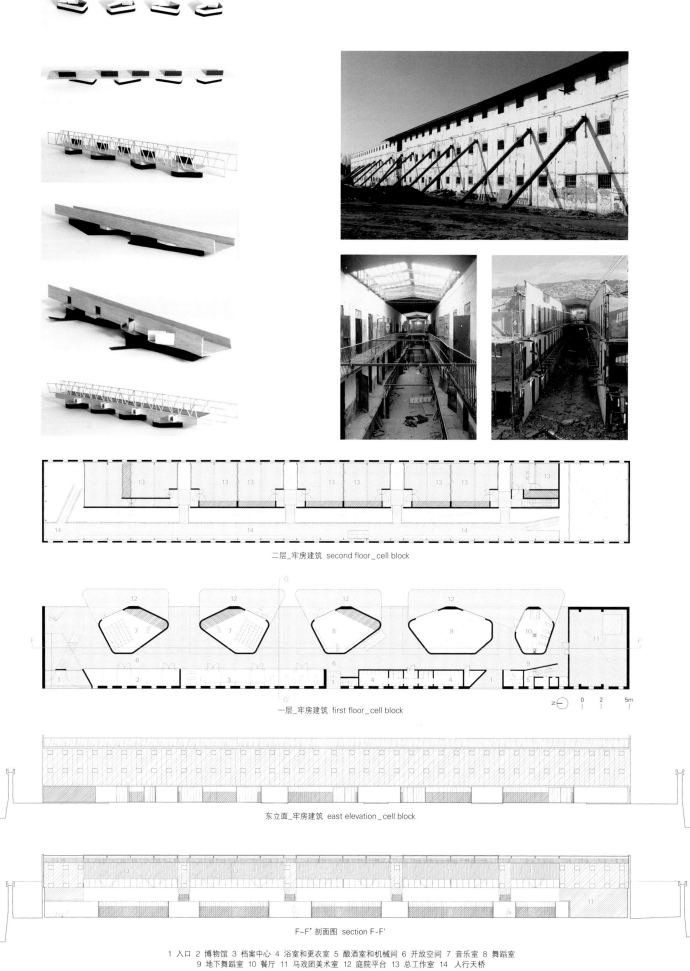

二层_牢房建筑 second floor_cell block

一层_牢房建筑 first floor_cell block

东立面_牢房建筑 east elevation_cell block

F-F' 剖面图 section F-F'

1 入口  2 博物馆  3 档案中心  4 浴室和更衣室  5 酿酒室和机械间  6 开放空间  7 音乐室  8 舞蹈室
9 地下舞蹈室  10 餐厅  11 马戏团美术室  12 庭院平台  13 总工作室  14 人行天桥

1. entrance  2. museum  3. documentation center  4. bathrooms and changing rooms  5. winery and machine rooms  6. open space  7. music room  8. dance room
9. underground dance room  10. dining room  11. circus arts room  12. courtyard terraces  13. general workshop  14. skywalk

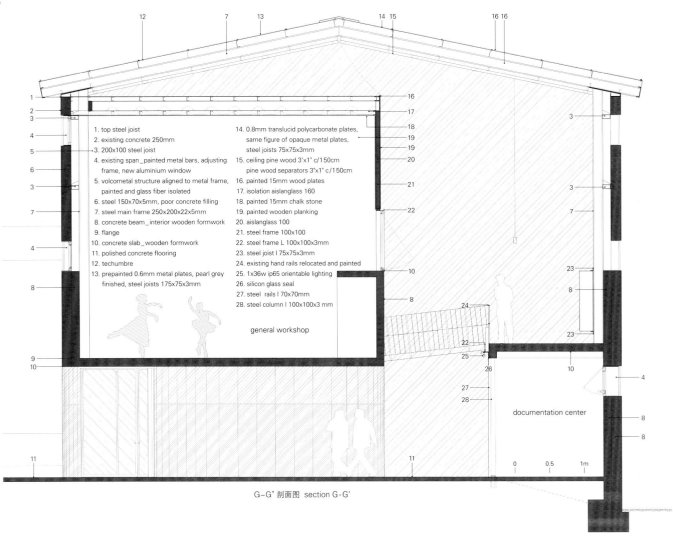

1. top steel joist
2. existing concrete 250mm
3. 200x100 steel joist
4. existing span _ painted metal bars, adjusting frame, new aluminium window
5. volcometal structure aligned to metal frame, painted and glass fiber isolated
6. steel 150x70x5mm, poor concrete filling
7. steel main frame 250x200x22x5mm
8. concrete beam _ interior wooden formwork
9. flange
10. concrete slab _ wooden formwork
11. polished concrete flooring
12. techumbre
13. prepainted 0.6mm metal plates, pearl grey finished, steel joists 175x75x3mm
14. 0.8mm translucid polycarbonate plates, same figure of opaque metal plates, steel joists 75x75x3mm
15. ceiling pine wood 3"x1" c/150cm pine wood separators 3"x1" c/150cm
16. painted 15mm wood plates
17. isolation aislanglass 160
18. painted 15mm chalk stone
19. painted wooden planking
20. aislanglass 100
21. steel frame 100x100
22. steel frame L 100x100x3mm
23. steel joist l 75x75x3mm
24. existing hand rails relocated and painted
25. 1x36w ip65 orientable lighting
26. silicon glass seal
27. steel rails l 70x70mm
28. steel column l 100x100x3 mm

general workshop

documentation center

G-G' 剖面图  section G-G'

# 景观实验室
## Cannata & Fernandes Arquitectos

该建筑位于历史之城吉马良斯中心的西南部Veiga de Creixomil。根据吉马良斯城市的总体规划,这里作为国家生态保护区,是一处景观价值非常高的敏感区域。

场地地形的特色为两侧标高差异较大。一侧是水道,一边是泥路,周围视野开阔,并且附近有大片绿色景观。

人们在高速公路上就可以看到这一建筑,隐约地看见一系列的嵌入结构,确保它能在这一区域与其他部分分开。

通往实验室的路在东面,是由拆除的小屋顶铺砌而成。这些小屋顶位于现存建筑物与相邻场地之间的通道之上。整座建筑物由接待广场和停车场隔开,它们与小路共同引导游客或使用者通过散步道到达主入口,并且和建筑最令人印象深刻的部分,Ribeira de Selho水域一侧的立面联系起来。

建筑内部主要的设计目的是突出工厂式的空间特点。通过简单却不失严谨的复原设计,这处明亮开阔的空间随时可用于新用途,或是满足各种功能项目的需求。

不同的功能模块的分割遵循着屋顶的结构,以在带有之前结构的不同空间和特色空间之间保持一种平衡。

建筑东侧是核心技术区,设有锅炉、UTAN及储水设施。

而在西侧,上层的现存体量之前处于废弃状态,也没有任何施工,现在都进行了重新布局,或是重建。

该提案旨在恢复原有建筑的特点的同时,也建造了一个新的屋顶,并对体量进行重建。毫无疑问,这座重建建筑具有修复建筑的当代特质。

增建结构的设计理念旨在明确突出和准确识别各个增建阶段,而不会产生对建筑历史的淡化及曲解。因此,建筑师对砌石中的构件进行了回收利用、清洗和更换。原有的砌砖或是其他材料(已高度腐烂)建成的体量与新体量一起采用白色混凝土来建造。不同时代和用途对不同材料有着不同的解读。

## Landscape Laboratory

The building is located in Veiga de Creixomil southwest from the historic city center of Guimarães, in a sensitive area of high landscape value, classified in accordance with the Guimarães Municipal Master Plan, as National Ecological Reserve.

The morphology of the terrain is characterized by a relative difference of height, by the presence of a water channel, dirt trails, great visibility from surrounding areas and close proximity of large green masses.

The visibility that the building has from the motorway implied a series of interventions, guaranteeing that this is an element which distinguishes itself in the territory.

The access to the Laboratory, on the East, created by the demolition of small roofs, which were on the passageway located between the existing building and the neighboring plot, and the separation of the building by the creation of a reception square and car park, lead the visitor/user through a promenade to the main entrance in contact with the most impressive moment of

北立面_旧建筑 north elevation_old

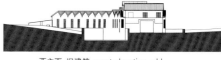

西立面_旧建筑 west elevation_old

北立面_新建筑 north elevation_new

西立面_新建筑 west elevation_new

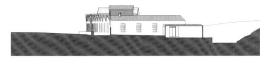

东立面_旧建筑 east elevation_old

南立面_旧建筑 south elevation_old

东立面_新建筑 west elevation_new

南立面_新建筑 south elevation_new

0  5  10m

©Pepe Barbiere (courtesy of the architect)

项目名称：Landscape Laboratory
地点：Guimarães, Portugal
建筑师：Fátima Fernandes, Michele Cannatà
项目团队：Riccardo Cannatà, Dario Cannatà,
Bruno Silva, Marta Lemos, Nuno Castro,
Francisco Meireles, João Pedro Martins
结构工程师：João Maria Sobreira
水力工程师：Raquel Fernandes
机械工程师：Raul Bessa
电气工程师：Alexandre Martins
音响工程师：Octávio Inácio
景观建筑师：Luís Guedes de Carvalho
施工单位：Construções Imobiliárias e Turísticas, S.A.
甲方：Câmara Municipal de Guimarães
施工面积：1,387.03m² 用地面积：988.98m²
总建筑面积：1,120m² 有效楼层面积：1,320m²
造价：EUR 1,108,000 设计时间：2010 竣工时间：2012
摄影师：
©Luis Ferreira Alves(courtesy of the architect) - p.83, p.84,
p.85, p.86, p.89, p.90, p.91, p.92, p.93
©Dario Cannatà(courtesy of the architect) - p.82,
p.88(except as noted)

A-A'剖面图 section A-A'

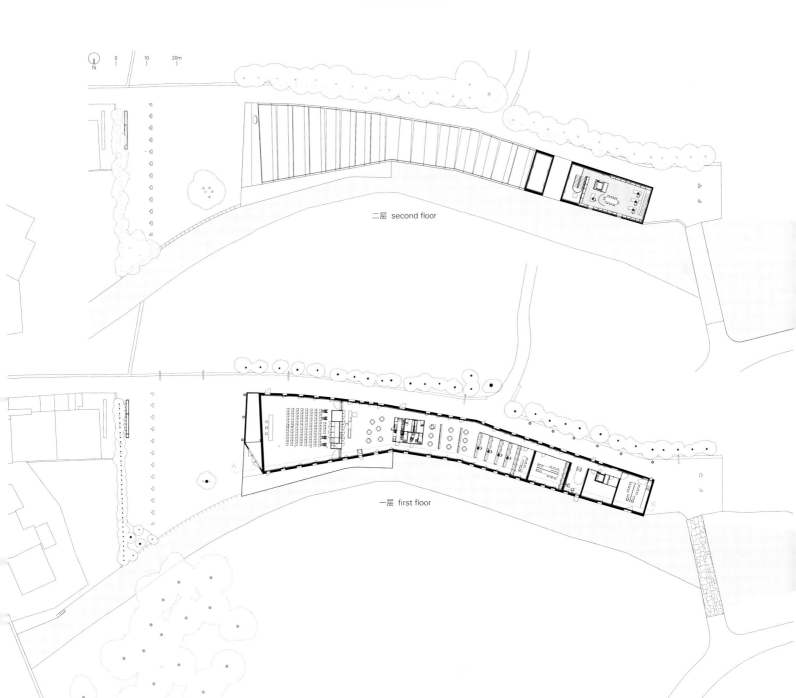

二层 second floor

一层 first floor

the building: the facade to the water front Ribeira de Selho.

On the inside, the main objective was to enhance the spatial character of the factory typology. An open space, bright, is ready to embrace the new uses and demands of it's functional program, through a simple but rigorous rehabilitation.

The division of the different functional modules follows the structure of the roofs in order to establish a balance between the different spaces with the structure of the preexistence, and especially with its character.

On the East is located the necessary technical area for the boiler, UTAN and water deposits.

On the West, the existing volumes on the upper floor, that were in a state of ruin or without architectural or construction quality were rearranged and reconstructed.

The proposal aims to recover the architectural character of the existing building and at the same time, with the new roof and the volumetric reconstructions, affirming the indisputable contemporaneity of a rehabilitated building.

The concept of intervention seeks to clearly highlight and accurately identify the various intervention stages so as not to produce ambiguities or distortions of the history of the building. Thus the elements in masonry stone were recovered, cleaned and replaced, the existing volumes in brick, or other materials in an advanced state of decay and new volumes were build in white concrete. Different materials allow a reading of changes in time and use.

详图3 detail 3

1. existing stone wall
2. 70mm metal profile
3. thermal insulation 90mm
4. masonry wall 110mm
5. 10mm galvanized steel sheet
6. double foil laminated gypsum, type "knauf" 12.5mm
7. galvanized steel frame, painted white 10mm
8. double glazing type "solar glass" supernatural solarlux 52/55 ht 8mm templex room, 18mm
9. thickness
10. perforated for ventilation grille in galvanized steel painted white, 10mm
    space for ventilation (intake)

1. concrete roof with a helicopter finishing
2. inner tube camera d'aria
3. vapor barrier
4. thermal insulation
5. mineral wool panels type "baswaphon", calibrated and with reduced weight
6. paint of all gypsum boards surfaces with water paint type "sika passimur", equivalent to apply with 3 coats, color: white mate
7. M22
8. kicking in wood freijo or teak, lacquering with white colour
9. floor in American treaty oakwood with 20mm of thickness, composed by slats with 12cm of minimum width, length 2.5m
10. tube type "wirsbo evalpex 16x1.8"
11. screed with additive
12. isolation
13. concrete slab
14. fixing profile of laminated gypsum type "knauff"
15. acoustic insulation in rock wool thickness: 40mm and 40kg/m³ of density
16. black veil
17. 2 laminated gypsum boards type "knauf" thickness of 12.5 mm +12.5 mm, metallic structure included in profiles of 48 mm
18. wall of concrete according to structures draft
19. cement brick wall
20. insulation 33mm
21. box brita
22. concrete slab with wire mesh type "socitrel cq38" thickness 15cm
23. compacted soil
24. PVC waterproof cloth and include a polyester armor, type "alkorplan®", "sika trocal, sgma"
25. IPE 80

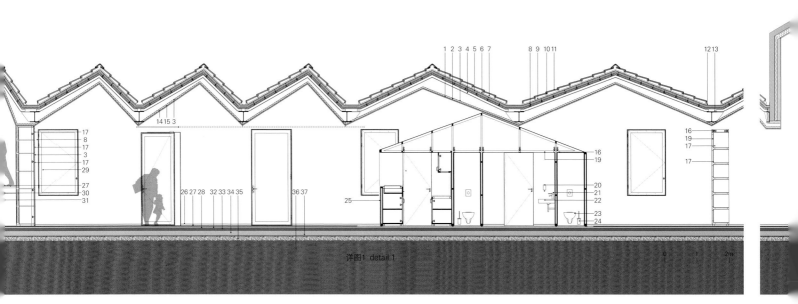

详图1 detail 1

26. rigid insulation board, polystyrene extruded type "roofmate" with the thickness 6cm
27. panel "sandwich" coated polyurethane type "erfi" system sub-tile 6cm of thickness
28. wood
29. profile of anchorage
30. marseille tile
31. eaves zinc in variable dimension
32. HEB 320
33. fluorescent lamp
34. panel lacquered MDF white colour thickness 10mm
35. structure in solid pine treated wood and pre-immunized acrylic opal
36. dispenser soap type "senda" ref. 001006 with chrome
37. cover and white polyethylene container

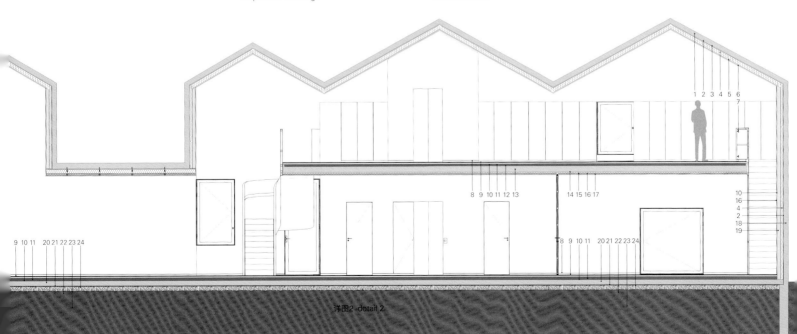

详图2-detail 2.

1. existing soil
2. yellow granite rail with exterior fine finish with 60cm of width
3. granite cube
4. layer of sand with cement particles in dry thickness. minimum 50mm
5. joint filled with sand
6. layer of laying precast concrete
7. tout venant (minimum thickness 15cm)
8. soil compacted
9. paint of all surfaces of plaster with water based paints like "sika passimur" or equivalent, to be applied with three coats of matte white color
10. two plasterboard laminated type "knauff" thick 125mm+125mm , including the metal structure in sections of 48mm
11. thermal insulation 90mm thickness
12. vapor barrier 13. inner tube
14. coverage of concrete, according to the design of the structure, 150mm
15. concrete roof with a helicopter finishing
16. concrete bricks wall
17. wall of concrete, according to the design of the structure, 150mm
18. galvanized steel railing thickness 10mm lacquered in white color
19. covering of the frame with stainless steel type "CBC"
20. grid for ventilation only in the rooms indicated in the design of hvac
21. wood panel of freijo or teak, lacquered in white color
22. floor in American oakwood treaty of 20mm thick, consisting of tables of minimum 120mm wide and 2.5m long
23. tube type evalpex 16x1.8"
24. screed with additive  25.
25. 33mm insulation
26. slab of concrete with mesh like "socitrel cq38 " thick 150mm
27. layer of gravel  28. drainage
29. cloth plasticized pvc waterproofing and integrated armor polyester fabric, such as "alkorplan®", "sika trocal, sgma"
30. crowning of corten steel thickness 10mm
31. recovery of integration with existing wall of a stone with the same characteristics of the existing

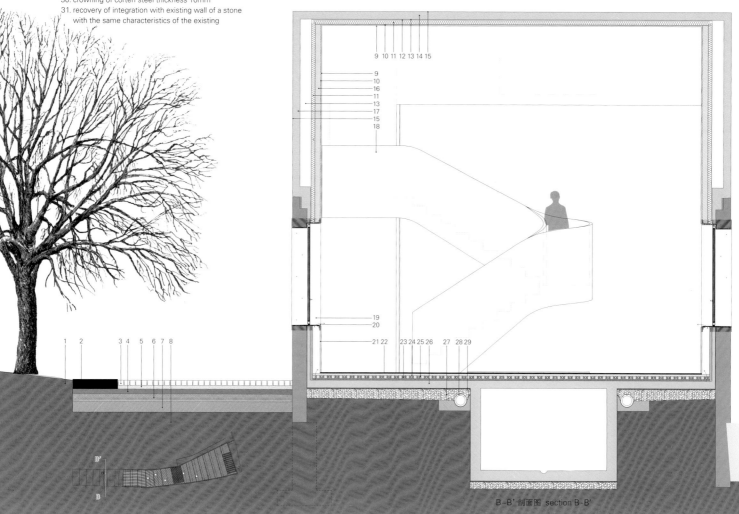

B-B'剖面图 section B-B'

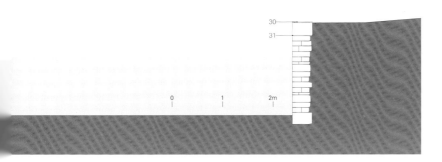

# Casa Mediterraneo总部
Manuel Ocaña

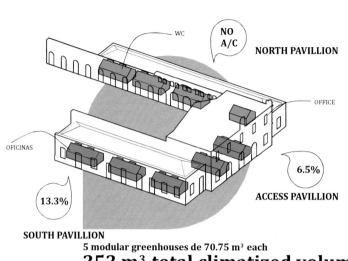

**5 modular greenhouses de 70.75 m³ each**
**353 m³ total climatized volume**

*we climatize people, we do not climatize spaces*

**旧贝纳卢亚火车站的翻新以及Casa Mediterraneo总部建筑的嵌入**

Casa Mediterraneo总部是一个外交机构,致力于培养地中海村的共同特征。由于该机构开展大范围的活动,如展览、音乐会、表演或聚会,需要新的空间,因此新总部设立在阿利坎特的旧贝纳卢亚火车站内。

除去项目要求,建筑师必须要解决下面隐含的问题:"牢记保护主义者和项目书要求低预算的前提,让我们的视野不仅仅放在对建筑进行的简单防腐处理方面。我们应尝试设计一个彻底的却亲民的嵌入结构。那么,我们应该如何处理建筑遗产呢?"

首先,当建筑师正处于所谓的翻新模式时,他们应该摒弃现代主义过剩的荷尔蒙。其次,更新并添加朴素及特殊的装饰。最重要的是,将之前用于轨道交通的大型线性空间精细化,使之成为一处公共的体验空间。

因此,为了突出该建筑有价值的特性,建筑师对其进行了预先观测。经过探测,一些过剩的东西被彻底去除。因此,嵌入的结构就可以专注于将新属性融入内部空间。

之前的平台大厅是该建筑的主要空间。建筑师旨在将一处1500m²的纵向空间改建为一个体验大厅。通过无处不在的地中海阳光和一些额外的技术干预措施——例如悬挂各种物件,包括ø7m吊扇——使它的空间属性有所改善。一处未使用的、弥漫着锈气的、黑暗且干燥的空间呈现出蓝色,使人兴奋,具有流动感,充满活力,不死板,如河水般充沛,设有天顶,有节奏感,呈现出程序化模式,使用了热动力并且可获益。

一个克莱因蓝色的半透明屋顶滤进阳光,光线通过一个充满活力的环形铝质格架反射后,使旧墙面和地面有了颜色并且富有激情,并且充斥着整个空间,进而转化为蓝色的海洋。

一排小展馆被分散开来,沿着长廊设置,以用于一些传统的功能。这些展馆设备完善,并且可以调节温度。这些可使任何工人或参观者感受并理解建筑最初的结构和空间。建筑其余部分的温度是不可调节的。尽管有所遮盖,但空间并不是封闭的,新鲜空气可以源源不断地进入。地板材质为精简砂,而园艺呈现为光滑的陶瓷盆布局。

设计的结果呈现了一个遗产建筑的全新改建草案。在这个新草案中,旧的建筑物与公民会产生共鸣,在感受记忆的同时,可以避免任何为怀旧而做出的让步。

## Casa Mediterraneo Headquarters

### Refurbishment of the Old Benalúa Station and Insertion of Casa Mediterraneo Headquarters

Casa Mediterraneo is a diplomatic institution committed to foster the Mediterranean Villages' common identity. Its new Headquarters are to be set within the old Benalúa Railway Station, in Alicante, because the institution demanded new spaces where to develop a wide range of events, including exhibitions, concerts, shows or parties.

Beyond the programmatic requirements, the architects could not avoid tackling the following implicit issues: "Bearing in mind the preservationist and low budget premises established by the brief, setting our sights beyond the simple embalment of the build-

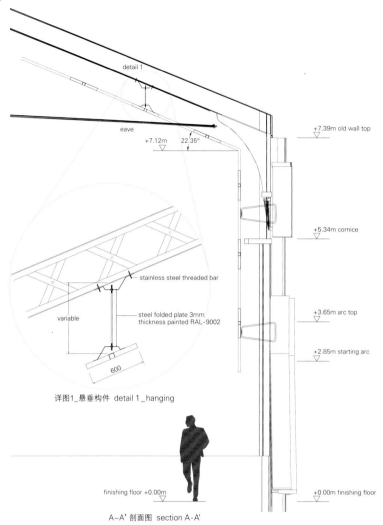

详图1_悬垂构件 detail 1_hanging

A-A' 剖面图 section A-A'

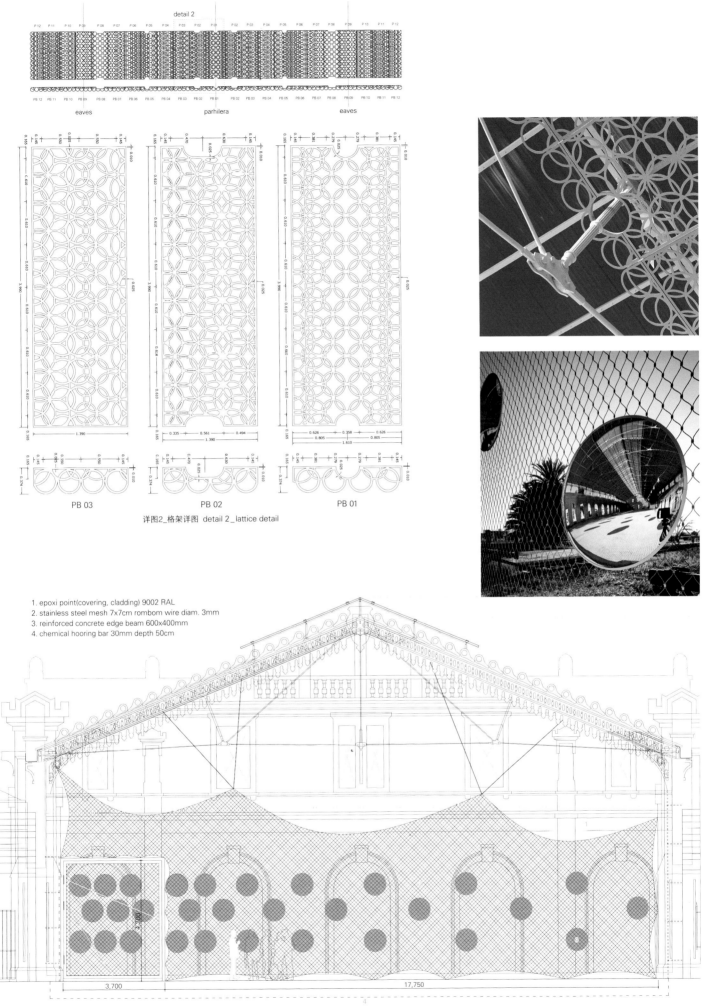

详图2_格架详图 detail 2_lattice detail

1. epoxi point(covering, cladding) 9002 RAL
2. stainless steel mesh 7x7cm rombom wire diam. 3mm
3. reinforced concrete edge beam 600x400mm
4. chemical hooring bar 30mm depth 50cm

西立面 west elevation

ing and attempting to carry out a radical-yet-citizen-friendly intervention, how were we to approach Heritage?'

Firstly, dismissing the Modernist excessive testosterone when operating into what could be described as Refurbishment Mode. Secondly, updating and adding straightforward and idiosyncratic ornamental references. And, above all, sophisticating that bold, giant-linear space – formerly devoted to the railway transit – and turning it into a public, experiential space.

Thus, and after pre-scanning the building in order to highlight its valuable features, as well as to detect and get rid of the fat surplus once and for all, the intervention could focus on incorporating new properties to the interior spaces.

The former platform hall is the main space of the building. A 1500 sqm, longitudinal space is intended to become an experience hall. By means of the ubiquitous Mediterranean Sun and some additional technical interventions – such as the introduction of a variety of hanging elements, including a ø7m ceiling fan – its spatial properties evolve. An unused, rusty, dark and dry space becomes blue, excited, liquid, vibrant, unstable, overflowing, zenithal, rythmic, programmable, thermodynamic and profitable.

A Klein-blue, translucent roof filters the incoming sunlight which, after reverberating through a vibrant, circular-patterned aluminium lattice, dyes and excites the old walls and floor, flooding the space and turning it into a sea of blue shadows.

The conventional programme is fitted into an array of small pavillions scattered along the perimetral aisles. These are equipped and climatized, and enable any eventual worker or onlooker to enjoy and understand the original structure and space. The rest of the building is not climatized. Though sheltered, the spaces are not closed, allowing a constant flow of fresh air inside. The floors are cast of compact sand, and the gardening is distributed in glossy ceramic pots.

The result is a new occupation and transformation protocol for heritage buildings. A new scenario in which the old buildings and the citizens empathize, feeds the flame of memory while avoiding any nostalgic concession.

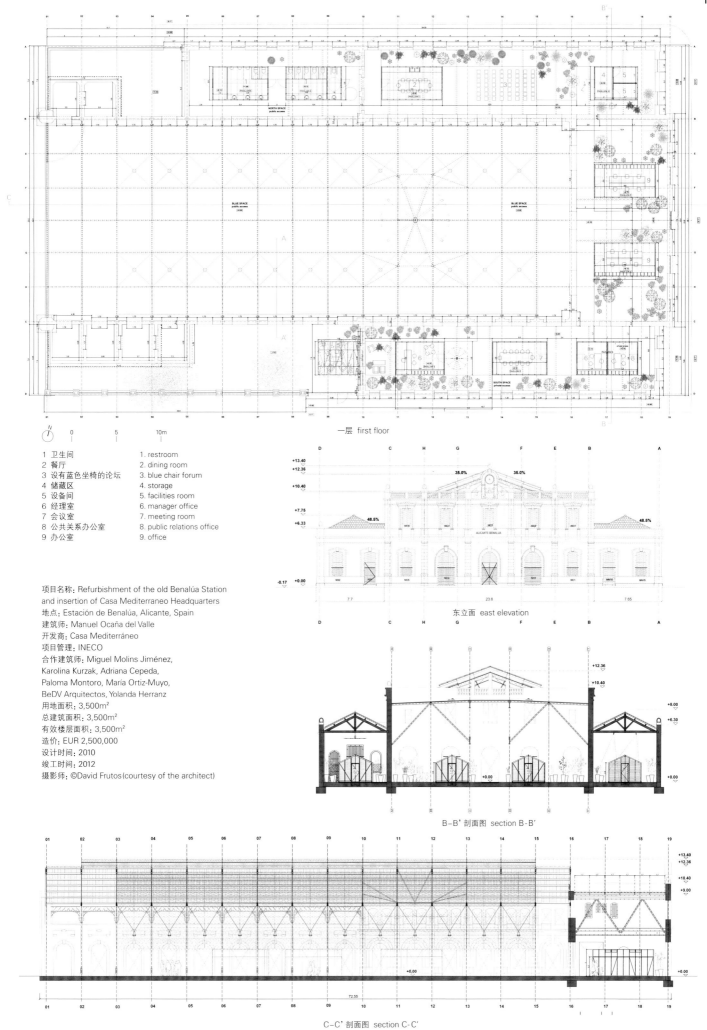

104

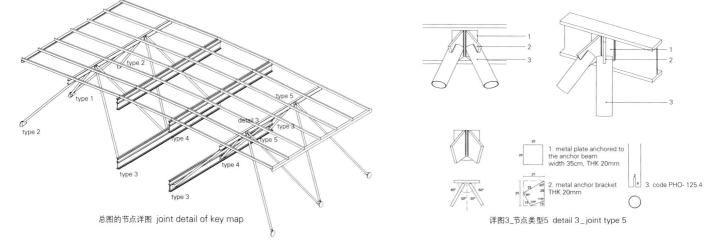

总图的节点详图  joint detail of key map

详图3_节点类型5  detail 3_ joint type 5

1. metal plate anchored to the anchor beam width 35cm, THK 20mm
2. metal anchor bracket THK 20mm
3. code PHO- 125.4

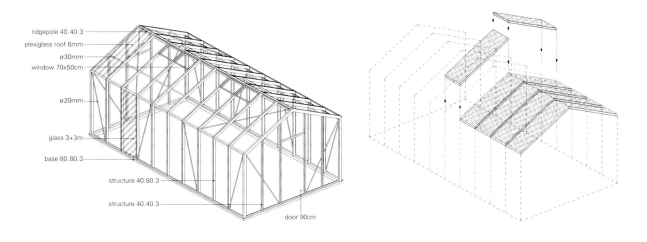

新世界 New Reality

## Conde Duque建筑
Carlos de Riaño Lozano

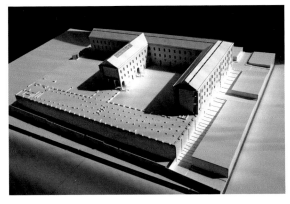

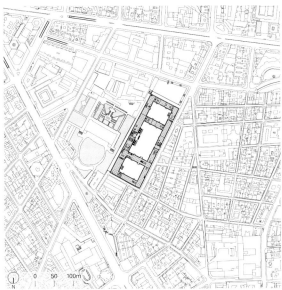

1717年，Pedro de Ribera在马德里绘制了Conde Duque军营的第一张设计图。这是一个有宏伟目标的项目，必须容纳超过600名的皇家卫队士兵及400匹的马。

近三个世纪以来，历经无数磨砺，该建筑一直保持着相同的结构。19世纪下半叶的两起火灾，算是这座建筑所经历的最大的磨难，几乎使它需要拆除并进行重建，而修复工作至今仍未完成。

2004年规划总监的设计草图给了Conde Duque建筑一次新的冒险。这个设计对整座建筑进行了整修，并于2011年完成。

由于它是马德里继旧皇宫阿卡扎堡之后最大的建筑，在当时，这算是一次冒险的赌博。至今，它仍然是这座城市最大、最少见的历史建筑之一。

建筑师对于它裸露的内部的第一印象是开放的中殿，带有砖砌拱顶，位于花岗岩柱子之上，以及其他楼层制作精细的柱子和铆接的横梁。实心砖结构形成干净且简朴的空间。

坚实的空间需要一个明朗的解决方案：保留框架，去除增建部分，集中建造新的结构。其目的是建造一个大型中性容器，能够容纳永恒的物件以及一些特别的元素。

提议了这个项目的马德里市政府非常清楚这些前提，并积极推进任务的完成。这个项目的本质目的是想创建互动的地方、多功能空间以及为一些活动发展而建的小型集合场地（一座音乐厅、一座剧院、一个礼堂……）。

内部的庭院形成了庞大且多元化的功能空间。这些庭院也能恢复开放的空间，并实现室内与室外之间的流动。对于外界而言，这座建筑还是非常密闭的。但是，这些庭院还是可以进入的，且给人一种亲和感。它们是接待室，一处可以接近的地方，一处可使人理解的地方。

增建设计包括两个阶段。第一阶段是在建筑的南区。原建筑在这里展现了它最真实的一面，这里也是供展览和表演的大厅所在。一排大房间通过小型百叶窗匣来改变其面貌，而小型百叶窗匣用来隐藏新的通信通道。一些独特的建筑，比如音乐厅、剧院及礼堂，都有自己的氛围及对环境的触动。

第二阶段是在北区。在过去的几年里，这里进行了大规模的改建。建筑师在没有大量自由发挥的基础上，重组了现有的机构，比如档案馆、报纸图书馆、图书馆、天文台及当代艺术博物馆。

整座建筑有着共同的语言。幸存下来的最初建筑结构被复原，在上面又涂了油漆，就像是在均匀连续的画布上作画。增建部分被漆成白色。而建筑的立面则述说着自己的历史。

如果它的立面，包括门楣和檐口，是决定这次修复的基础的话，那么空洞便是它最微妙的部分了。它们完美地排列着，以重复的节奏，恢复了垂直的比例。空洞上包覆木材和铁，为的是在不降低室内温度的同时，最大限度地减少来自外部的影响。

### Conde Duque

In 1717, Pedro de Ribera drew the first project of the Conde Duque Barracks in Madrid. It had to provide answers for an ambitious program to house more than 600 men and 400 horses of the Royal Guard.

For almost three centuries the building has maintained the same structure, having overcome countless trials. The most significant ones were two fires in the second half of the 19th Century that almost required its demolition and rebuilding. The restoration was never complete.

In 2004, with the drafting of the Plan Director, a new adventure in the life of the Conde Duque began, which finished with the completion of the works in 2011, after a total and coordinated renovation of the whole building.

In its moment it was a risky gamble, since it was the largest build-

ing in Madrid after the Alcázar, the old royal palace. And it is still one the hugest and less-known historic buildings of the city.

The first images of its naked interior were open naves of brick arches on granite pillars and at other levels, wrought columns and riveted beams. Clean and austere spaces are covered with solid brick structures.

The powerful space called for an obvious solution: to maintain the skeleton, to remove additions and to concentrate the new interventions. The aim was to achieve a big neutral container where new timeless objects and specific elements could be situated.

The program is proposed by the City of Madrid who was aware of those premises and facilitated the task. It essentially wanted to create places of interaction, multi-use spaces and a small collection of venues for the development of activities (a concert hall, a theatre, an assembly hall…).

The interior courtyards shape the vast and diverse program. They are also an opportunity to recover open spaces and to allow the indoors and outdoors flow. The building is still very hermetic towards the exterior, but these courts offer a friendly and accessible face. They mean the antechamber, the place of approach, the possibility of understanding.

The intervention consisted of two phases. The first one took place in the southern area of the building, where the original architecture shows the most authenticity, which houses halls for exhibitions and shows. A sequence of big rooms is only altered by small blind boxes that hide the new communication passages. The unique pieces, like the Concert Hall, the Theatre and the Assembly Hall, generate their own ambiance and give their own touch to each environment.

The second phase took place in the northern area, where the original container had been extensively transformed throughout the last years. It reorganized less freely the existing institutions, such as the Archives, the Newspaper Library, the Library, the Observatory and the Museum of Contemporary Art.

A common language unified the whole building. The original structures which had survived were recovered and liberated of superposed coats of paint, achieving a homogeneous and continuous canvas. The additions were painted white and the architects let the elevations tell them their own history.

If the surfaces, including lintels and cornices, are the base of this discourse, it is the hollows that provide the nuances. Perfectly ordered, with a repetitive rhythm, now it recovers the vertical proportion, and it is dressed with a mixed carpentry of iron and wood that minimizes the impact from the exterior without subtracting warmth to the interior.

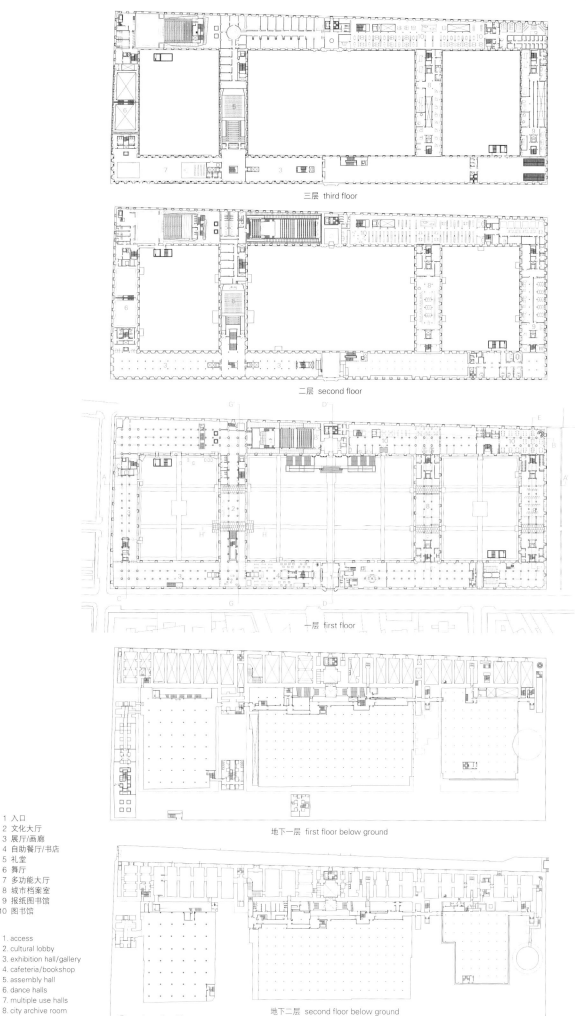

1 入口
2 文化大厅
3 展厅/画廊
4 自助餐厅/书店
5 礼堂
6 舞厅
7 多功能大厅
8 城市档案室
9 报纸图书馆
10 图书馆

1. access
2. cultural lobby
3. exhibition hall/gallery
4. cafeteria/bookshop
5. assembly hall
6. dance halls
7. multiple use halls
8. city archive room
9. newspaper library
10. library

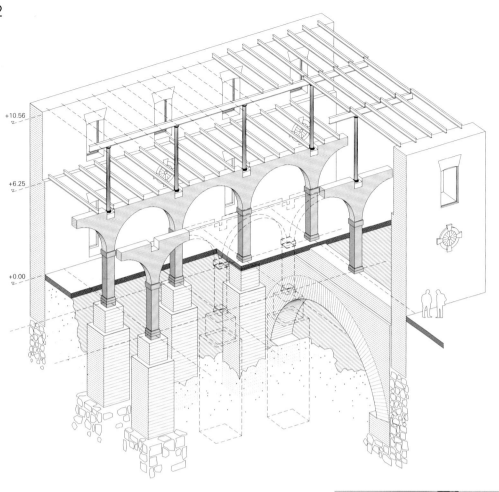

西北立面 north-west elevation

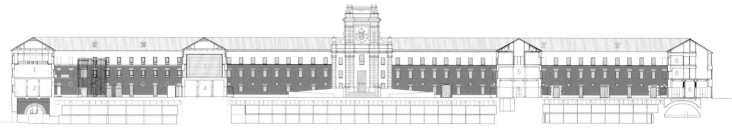

A-A' 剖面图 section A-A'

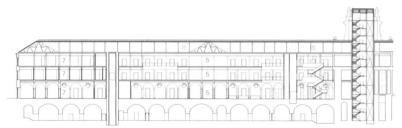

B-B' 剖面图 section B-B'

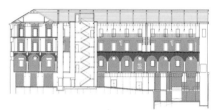

C-C' 剖面图 section C-C'

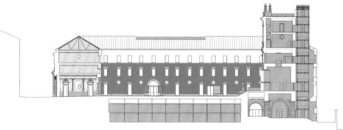

D-D' 剖面图 section D-D'

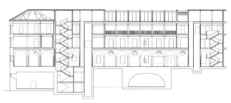

E-E' 剖面图 section E-E'

| | |
|---|---|
| 1 舞厅 | 1. dance halls |
| 2 展厅/画廊 | 2. exhibition hall/gallery |
| 3 礼堂 | 3. assembly hall |
| 4 文化大厅 | 4. cultural lobby |
| 5 城市档案室 | 5. city archive room |
| 6 报纸图书馆 | 6. newspaper library |
| 7 图书馆 | 7. library |
| 8 入口 | 8. access |
| 9 多功能大厅 | 9. multiple use halls |

F-F' 剖面图 section F-F'

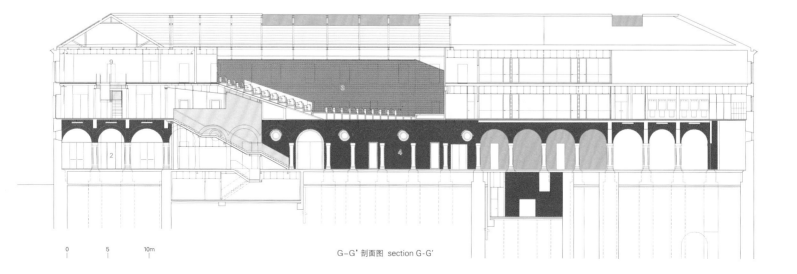

G-G' 剖面图 section G-G'

项目名称：Conde Duque
地点：Calle Conde Duque s/n, Madrid, Spain
建筑师：Carlos de Riaño Lozano
合作建筑师：Almudena Peralta Quintana, Rebeca Hurtado Diaz
技术建筑师：María del Hierro, Luis García Cebadera
工程设备：JG Instalaciones
结构工程师：Otep Internacional
照明工程师：Intervento
发起人：Ayuntamiento de Madrid, Área de Las Artes
施工单位：Dhinor 立面修复：CPA
礼堂+展厅+排练厅：EDHINOR
城市档案室：UTE Geocisa y Fernandez Molina
图书馆+天文台：UTE Geocisa y Fernandez Molina
城市报纸图书馆：PECSA
用地面积：19,390m²
总建筑面积：10,323m²
有效楼层面积：29,262m²
设计时间：2009
施工时间：2010
竣工时间：2011
摄影师：©Miguel de Guzman(courtesy of the architect)

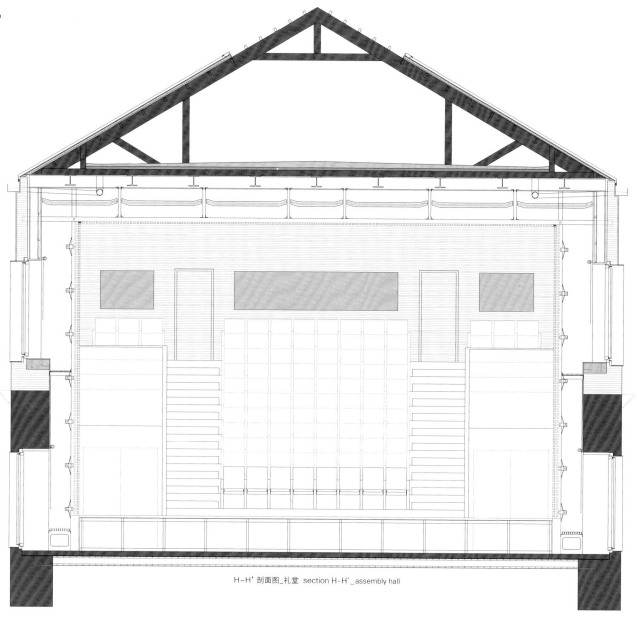

H-H'剖面图_礼堂 section H-H'_assembly hall

新世界 New Reality

# Can Ribas工厂的修复
Jaime J. Ferrer Forés

马略卡岛帕尔马的La Soledat区正在开展一项都市项目，旨在复兴附近一带并保护工业遗留下来的元素。自2005年赢得公开竞争以来，此项提议作为城市总体发展、公共空间和工业遗留建筑的修复，以及新社会住房项目，正在逐步实施。

1851年，致力于生产毛毯的Can Ribas工厂建于城外。后来，在1945年，工厂及La Soledat区的外围核心被合并到了扩建区域。尽管如此，到了20世纪70年代，经历几番扩建及改建后，工厂变得荒废，最终被废弃了。与此同时，周围环境也在发生着重要变化：大型社会住房的建立以及1965年Son Molines发电站的兴建，这些创造了一个全新的障碍，进一步强化了这一地区的物理隔离。

接下来，试图通过阻止衰败进程达到重建该地区的尝试失败了。因此，在2003年，一项针对整个海滨地区的专项规划获得批准。该计划包括Levante及La Soledat附近地区，旨在采取不同行动来推进规划效果。因为它被规划成为一个封闭的管辖区，Can Ribas工厂已经成为一个将La Soledat一分为二的障碍。这也促成了2005年重新发展公共空间的一场竞赛：Can Ribas工厂主体建筑的翻新以及社会住房的建造。

第一阶段已经建立了一片围绕工厂主体建筑及烟囱的公共区域。它连接了两块空地并消除了费里奥尔街沿线的交通问题。此外，主要通道转化为建筑内的市民中心，并设在新广场。工厂及烟囱是仅有的列管建筑。出于这个原因，受到新街道影响的建筑已部分拆除。然而在项目开发过程中，最宝贵的遗留元素已经恢复，并将融入城市环境。建筑主体剩余部分里建造的都市门廊、蒸汽馆及其中一个仓库的立面，彰显了这个工业生产基地的价值。这个基地由纺织生产过程的不同阶段所需的场馆组成。公共开放空间的新布局是通过连接新Brotad街和被保留的工厂要素的混凝土基座形成的。石英岩搭配混凝土板铺筑的空间显得更加丰富且复杂。

都市门廊、蒸汽馆及另一个仓库墙面的结合，体现了工业区的价值。如此一个具有混凝土基础的公共开放空间体系，在Brotad街和Can Ribas工厂间建立了一个视觉和物理连接，创造出了更丰富、更复杂的公共空间。

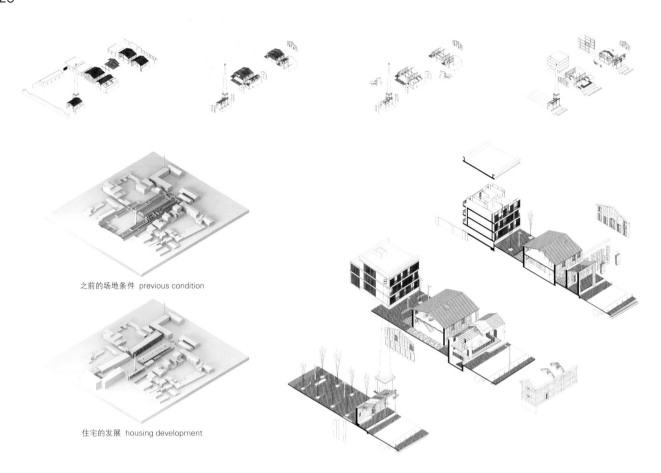

之前的场地条件 previous condition

住宅的发展 housing development

项目名称：Can Ribas
地点：Brotad Street, La Soledat, Palma de Mallorca, Spain
建筑师：Jaime J. Ferrer Forés
合作建筑师：project architect-Yolanda Ortega Sanz/
student-María Alemany Perelló/site supervision-Jaime J. Ferrer Forés/
models-Yolanda Ortega Sanz, Talleres Cortada
顾问：urban services-Antoni Ramis Arrom, Esteban Pisano Porada/
mechanical engineering-Jaime Ferrari/engineering-Inés Batle Eriksson/
heritage structure-Eduard Simó/heritage quantity surveyor-María Antonia
Palmer Poncell, Bartomeu Bonet Palmer
承包商：Melchor Mascaró, Bartolomé Ramón
工业遗产顾问：Toni Vilanova
甲方：Ajuntament de Palma, Consorci Riba,
Patronat Municipal de l'Habitatge
用地面积：5,257.42m²  总建筑面积：287.50m²
竞赛时间：2005  施工时间：2010—2011
摄影师：©José Hevia(courtesy of the architect)

## Can Ribas Factory Renovation

Neighborhood revitalization and preserving the elements of industrial heritage are the aims of this urban project being undertaken in the La Soledat area of Palma, Mallorca. Proposal is being developed in several stages after winning the open competition in 2005 and consisting in urban general development, public spaces, industrial heritage restoration and new social housing.

Dedicated to the production of wool blanket Can Ribas Factory was built in 1851 outside the city. Later on, in 1945, the factory and peripheral core of La Soledat would be merged into the expansion areas, although in the 1970s, after several extensions and transformations, the factory became obsolete and fell into disuse. At the same time important changes occurred in the surroundings: a large social housing development was built and the power station of Son Molines went up in 1965, creating a new barrier that further increased the quarter's physical isolation.

The later attempts to redevelop the area were unsuccessful in halting the process of decay, so in 2003 a special plan for the whole waterfront was approved. The plan included the Levante and La Soledat neighborhoods to boost the effect of the different actions undertaken. Since it was organized as a closed precinct, the factory of Can Ribas had become a barrier dividing La Soledat into two, and this encouraged to call a competition in 2005 for the redevelopment of the public spaces, the refurbishment of the main building of Can Ribas and the construction of social housing. The first phase has already established a public area around the main factory building and the chimney, connecting two empty spaces and eliminating traffic along Ferriol Street. Furthermore, it inverts the main access to the civic center in the building, placing it in the new plaza. The factory and the chimney were the only listed elements. For this reason the buildings affected by the new street have been partially demolished. In developing the project, however, the most valuable heritage elements have been recovered and will be integrated into the urban surroundings.

The creation of an urban porch in the remaining part of the main building, the steam pavilion and the facade of another one of the warehouses emphasize the value of this industrial complex that comprised several pavilions for the different stages of the textile manufacturing process. The new arrangement of the public open spaces is organized by a concrete plinth connecting new Brotad Street with the preserved elements of the factory, generating a richer and more complex space paved with bands of quartzite stone combined with cast concrete slabs.

The incorporation of the urban porch, the steam pavilion and the wall of another warehouse enables recognition of the value of the industrial area. A system of open public spaces is thus structured by a concrete foundation, which serves to create a visual and physical connection between the new Brotad Street and the historic elements of the Can Ribas factory generating a richer, more complex public space.

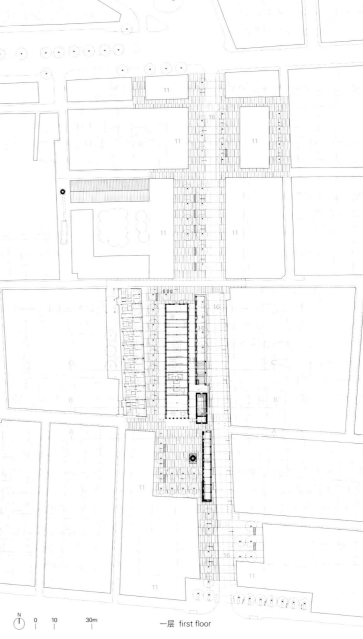

之前的纺织厂 former textile factory
第一阶段的开发 first stage development

| | | |
|---|---|---|
| 1 入口 | 6 场馆 | 12 主展馆、市民中心 |
| 2 接待处 | 7 主要的纺织品场馆 | 13 工业通道 |
| 3 庭院 | 8 仓库 | 14 展馆遗留部分的立面 |
| 4 拥有羊毛织物纹理和颜色的场馆 | 9 烟囱 | 15 都市门廊 |
| 5 蒸汽馆 | 10 纺织品饰面 | 16 新Brotad街 |
| | 11 新住宅建筑 | |

1. entrance
2. reception
3. courtyard
4. pavilion of fleece wash and tint
5. steam boiler pavilion
6. pavilion
7. main pavilion of spinning and textile
8. warehouse
9. chimney
10. textile finishing
11. new housing block
12. main pavilion, civic center
13. industrial passage
14. facade of the remaining part of the pavilion
15. urban porch
16. new Brotad Street

一层 first floor

南立面_Ferriol街 south elevation_Ferriol Steet

北立面_Ferriol街 north elevation_Ferriol Steet

西立面_工业通道 west elevation_industrial passage

东立面_Brotad街 east elevation_Brotad Street

0  10  20m

1 新住宅建筑
2 主展馆、市民中心
3 烟囱
4 蒸汽馆
5 都市门廊
6 新Brotad街
7 工业蒸汽炉原址
8 工业通道
9 场馆遗留部分的立面

1. new housing block
2. main pavilion, civic center
3. chimney
4. steam boiler pavilion
5. urban porch
6. new Brotad Street
7. industrial boiler premises
8. industrial passage
9. facade of the remaining part of the pavilion

A-A' 剖面图 section A-A'

B-B' 剖面图 section B-B'

C-C' 剖面图 section C-C'

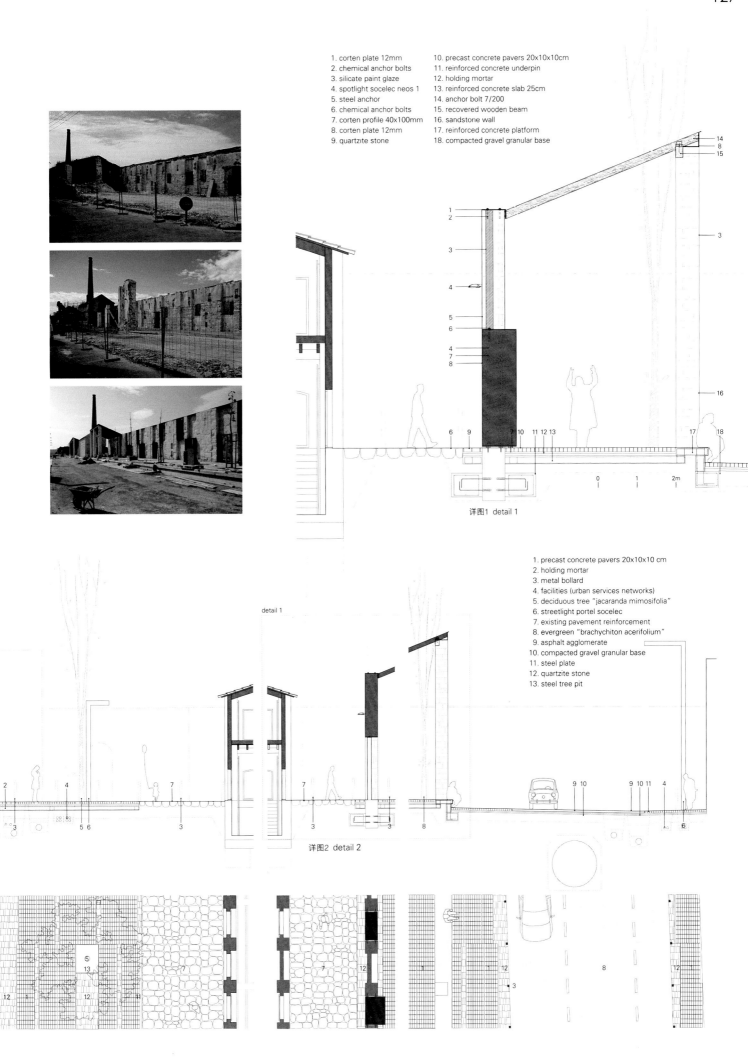

1. precast concrete pavers 20x10x10cm
2. holding mortar
3. quartzite stone
4. tubular profile ø10mm
5. steel railing reinforcement 60x90mm
6. steel railing 60x90mm
7. recovered wooden beam
8. L steel profile 60x60 mm
9. steel tube profile
10. corten steel plate
11. consolidation
12. sandstone wall
13. recovered tiles
14. waterproof microwave panel
15. waterproof plywood 12+12mm
16. wooden beam
17. wooden truss beam
18. lime plaster
19. silicate paint glaze
20. wooden framing
21. tempered glass 6+6
22. door handle erkoch
23. recovered wooden beam
24. chemical anchor bolts
25. steel plate
26. reinforced concrete slab 25cm
27. reinforced concrete platform
28. compacted gravel granular base

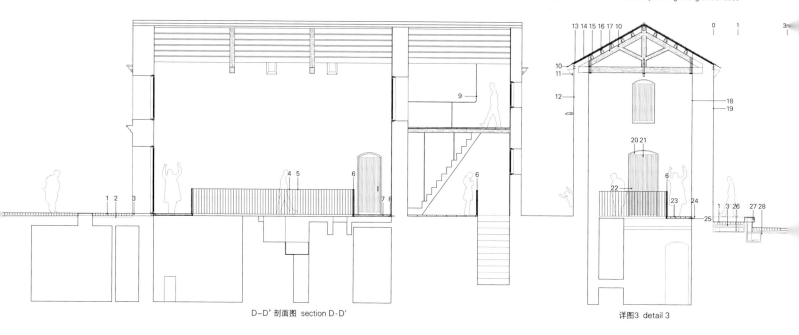

D-D' 剖面图 section D-D'        详图3 detail 3

1. corten steel profile
2. cut beam head
3. wooden beam
4. silicate paint glaze
5. existing sandstone wall
6. precast concrete pavers 20x10x10cm
7. holding mortar
8. reinforced concrete slab 25cm
9. compacted gravel granular base
10. reinforced concrete platform
11. metal plate
12. reinforced concrete pillar with embedded sandstone
13. existing limestone
14. waterproof plywood 12+12mm
15. waterproof microwave panel
16. recovered tiles
17. tempered glass 6+6
18. waterproofing
19. anchor bolt 7/200
20. corten steel plate
21. corten steel sheep
22. existing reinforcement brickwork
23. existing foundry pillar
24. corten steel framework 10m
25. chemical anchor bolts

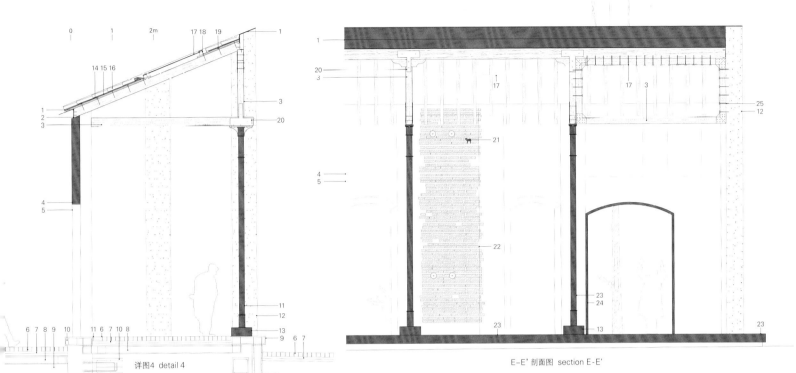

详图4 detail 4

E-E' 剖面图 section E-E'

# 彰显自由与功能的大学建筑类型学
# Liberal and Functional
## University Faculty Building Typology

在近代，出现了一种超越传统的、代表权力和宗教的纪念碑型新型公共建筑。这一类型建筑与知识和科学研究相关联，反映出自己独特的纪念碑性品牌。从大约12世纪第一批大学的诞生，到大约15世纪出现的第一批专门用于大学教育目的的建筑，再到20世纪下半页，大学建筑类型学这一定义产生了。虽然那时的大学建筑设计使用的仍然是当代建筑语言，但其所取得的成就可以称为经典。尽管这一类型的建筑实例多种多样，但这一固定下来的建筑类型的基本元素包括强大的功能整合、对建筑地点与学习和研究有关的环境条件的极大关注，以及与在大学社区内有助于促成充满活力的人际关系的环境和社会的密切联系。学术工作区、实验室、致力于放松身心的学生公寓和社交空间、运动和休闲设施，所有一切都为了促进知识的交换，来提高学术界的成果。在由路易斯·康设计的索尔克研究所和坎迪里斯设计的柏林自由大学项目这些现在称之为经典的例子之后，大学和研究机构的建筑设计语言旨在打破与世隔绝、唯我精英的建筑特点，相比之下，强调交流沟通性，达到室内工作空间与户外社会生活空间之间真正的渗透性。

Beyond the traditional monumentality linked to power and religion, a new kind of public building, associated with knowledge and scientific research and reflecting its own brand of monumentality, has emerged in more recent times. From the birth of the first universities around the XII Century and the first buildings dedicated specifically to that purpose around the XV Century, such construction in the second half of the XX Century arrived at a definition of a typology for university buildings which, although articulated in contemporary architectural language, achieved a state that could be called classical. The basic elements of this well-established typology, despite the wide variety of examples, are strong functional integration, significant attention to the environmental conditions of areas related to learning and research, and a close relationship with the environment that fosters a vibrant system of relationships within the university community and with the society. Academic workspaces, laboratories, residences and social spaces devoted to relaxation, sports and leisure produce a system of relationships that enhances the exchange of knowledge and the productivity of the academic community. After now the classic examples of the Salk Institute by Louis Kahn and the Freie Universität by Candilis, the architectural language of the university and research buildings has aimed at breaking a certain insulated, elitist character so as to communicate, by contrast, a real permeability between interior spaces dedicated to work and outdoor areas dedicated to social life.

### New Monuments to Knowledge

In the beginning architectural beauty was reserved to monuments of religion and power and to the cult of the dead. Such beauty bestowed a sacred value which was much more than hedonistic, and which was permitted only as a collective value, to which ordinary people had no access.

The departure from this state of affairs was introduced by the two civilizations which were the first to consider knowledge not as an esoteric heritage of dominant castes, but as a common heritage worthy of diffusion and celebration. Greek and Roman public places, the essential elements of the magnificentia civilis, were secular spaces dedicated, unquestionably, to trade and politics, but in no small measure, to the dissemination and discussion of knowledge as well. The Greek stoa, the Roman basilica and, last but not least, the thermal baths were the places where the origins of modern philosophical and scientific thought were founded. In these places the desire for free knowledge and for architectural beauty found their first conjunction.

The appearance in the XII century of the first public universities gradually led to the emergence of the first buildings devoted to research and teaching. Around the XV century, university buildings evolved into specialized typologies, providing the first functional spaces for scientific research and offering opportunities to create communities of teachers and students living around the places in which such knowledge was cultivated.

From the exclusive geographical centrality of the first universities, in the following centuries the model was perfected and disseminated on a large territorial scale, burgeoning into a network of universities and faculties that now connects all continents.

Thus, the university building typology has continued to produce examples of original and high architectural value, whose interest is enhanced by the fact that the great complexity and variety of hosted functions have produced an equally wide variety of shapes and designs, becoming a part, albeit for strictly functional reasons, of the modern architectural thinking that has lost interest in a semantics exclusively linked to function. The design of university buildings is executed on a case-by-case basis, according to a program related to specific use, rather than, as for the buildings of the civil magnificence of the past, to stylistic elements that allowed immediate recognition of the hosted function.

A great deal of ground was covered by the first buildings thus dedicated, such as the Archiginnasio of the old university of Bologna, via the concept of the Anglo-Saxon "urban" campus, to the great examples of modern university architecture. In the second half of the XX century, the typology of university buildings, albeit through the free and anticanonical language of the Modern Movement, achieved classical status.

The great examples of functional integration, high quality of life and deep inclusion in the context represent milestones of architectural thinking in general. First of all, though not precisely a university faculty, is the Salk Institute in La Jolla[1], designed by

RGS艺术、建筑和设计中心/Tadao Ando Architect & Associates
圣豪尔赫大学卫生学院/Taller Básico de Arquitectura
利默里克大学医学院和学生公寓/Grafton Architects
维也纳对外经济贸易大学的法学院和行政大楼/CRAB Studio

新的知识纪念碑/Aldo Vanini

Roberto Garza Sada Center for Arts, Architecture and Design/Tadao Ando Architect & Associates
Health Faculty of San Jorge University/Taller Básico de Arquitectura
Medical School, Student Residences at the University of Limerick/Grafton Architects
Vienna University's Law and Administration Buildings/CRAB Studio
New Monuments to Knowledge/Aldo Vanini

## 新的知识纪念碑

最初，建筑的美丽体现在宗教与权力的纪念碑，体现在对死者的崇拜方面。人们赋予这样的美一种神圣的价值。这种价值远远超过享乐价值，并只可作为集体价值，普通人达不到这样的价值。

最早与上述建筑美学和价值背道而驰的是希腊和罗马两大文明。这两大文明首次认为知识是值得传播和颂扬的共同遗产，而不是统治阶层秘传和私有的遗产。希腊和罗马的公共场所是这两大光辉文明的基本构成要素，毫无疑问也是人们长期从事贸易和讨论政治的地方，但在很大程度上，也是传播和讨论知识的地方。希腊的柱廊、罗马的长方形基督教堂，还有温泉浴室，这些都是现代哲学和科学思想建立的发源地。在这些地方，人们找到了对自由知识的渴望和对建筑美的渴望的一个结合点。

12世纪第一批公立大学的出现逐渐导致了专门用于研究和教学的建筑的兴起。大约到了12世纪，大学建筑逐渐演变，发展成为专门的类型学，并且首次为科学研究提供功能空间，传播知识，在增进知识的地方的周围创新教师和学生生活社区。

从最初大学所独有的地理向心性开始，在接下来的几个世纪中，这种大学模式得以完善，大学在更大规模的区域内扩建拓展，现在迅速发展成为连接各大洲的大学和学部网络。

因此，大学建筑类型学持续诞生具有独创性和建筑价值很高的建筑设计实例。这些建筑设计重视这样一个事实，即其功能的复杂性和多种多样性，同时也能够产生同等的形状和设计的复杂性和多种多样性。尽管仍受到严格的功能主义因素的影响，但这些建筑设计已经成为现代建筑设计理念的一部分，对只注重功能的语义论已经失去了兴趣。大学建筑的设计是基于个案基础之上的。每个项目都是根据具体用途而设计，而不是像过去那些富丽堂皇的建筑，注重的是风格鲜明的视觉要素，一眼就能看出建筑的功能。

根据具体用途而设计的大学建筑往往占地面积很大，根据盎格鲁-撒克逊的"城市"校园理念设计的古老的博洛尼亚大学所在地阿奇吉纳西欧宫就是其中之一，还有许多现代大学建筑的设计也是如此。在20世纪下半页，尽管大学建筑类型学经过崇尚自由、反对循规蹈矩的建筑语言的现代运动，但仍取得了经典这一份地位。

一般而言，一些整合功能、体现高品质生活、与环境深深契合的建筑设计案例代表并成为大学建筑类型学设计思想的里程碑。首先提到的就是由路易斯•康在20世纪60年代早期设计的位于拉荷亚的索尔克研究所[1]。虽然确切地说索尔克研究所不能算是大学建筑，但其明确的设计理念就是创造一处高质量的科学研究空间，正式性和功能性两大品质是这一高质量建筑的关键要素。第二个是坎迪里斯-约西齐-伍兹事务所为柏林自由大学设计的"低层大楼"[2]，真正成为了"城市的一部分"，是1963年至1973年之间实际施工在建的绝妙的建筑尝试，根据CIAM（国际现代建筑协会）确立的原则，完全是一个社会和文化的乌托邦项目。第三个建筑案例是莱斯特大学的工程学院项目[3]，由詹姆斯•斯特林和詹姆斯•高恩设计。两人在1964年确立了独创

Louis Kahn in the early sixties with the specific idea of creating a space where formal and functional qualities were a key element of the quality of scientific research. Second is the "groundscraper" of Candilis-Josic-Woods, for the Freie Universität of Berlin[2], an authentic "part of the city", a brilliant experiment in real construction, between 1963 and 1973, a project of social and cultural utopia according to principles established by the CIAM. Third is the Faculty of Engineering of the University of Leicester[3], by James Stirling and James Gowan, who established in 1964 an original and specific language, a new semantics of technological function: these are three examples among the many that make this typology one of contemporary architecture's areas of greatest experimentation. Despite its rejection of the canons of classical monumentality, the form has nevertheless created its own brand of monumentality linked to technological solutions related to the material and immaterial well-being of users. The need for proper lighting, energy efficiency, wide spaces free from structural impediments, and permeability to the urban context produces, in the best examples, a complex and varied repertoire of architectural responses.

The key concepts of the program design of these buildings include integration and liveability. The complexity of the functions is available to be fully experienced by the entire community of teachers, students, researchers and administrative staff, who have requirements ranging from privacy for individual study to the ability to gather large numbers of people, from the efficiency of technological laboratories to access to printed and digitized documents. A university faculty, however, is not merely a place of work, so the concept of liveability extends from purely technical characteristics to the definition of places for encounters that foster cultural and social exchange.

Spaces for relaxing and dining thus assume a central role. A considerable portion of the new ideas of contemporary knowledge is engendered in exchanges taking place in the cafeterias of university faculties, as well in the laboratories and in the academic spaces. In the examples of higher complexity, integration and liveability go so far as to accommodate housing and facilities for sports and entertainment.

As with other categories of contemporary architecture, special attention must always be paid to the creation of a permeability to contexts that can be urban areas, university communities, or in some cases, more or less populated countrysides. As compared to the traditional concept of a container with the openings and closures required for access and lighting, the building dismantles itself, and is carved and emptied as necessary to form a continuum with the outdoor spaces, such as a public road overlooked by the school's many academic and social functions.

Tadao Ando addresses these pursuits without sacrificing a strictly tectonic character in the Roberto Garza Sada Center of the University of Monterrey in Mexico. The quite regular concrete block is affected inside by virtual forces that empty and twist the remaining

维也纳大学法学院和中央行政大楼采用了富有曲线的、有机的、非正交的形态结构
Curvaceous, organic and non-orthogonal morphology implemented in Vienna University's Law and Administration Buildings

而特定的建筑语言，一种新的技术功能语义论。上面所述只是众多使大学建筑类型学成为当代建筑领域最伟大实验之一的三个例子。

尽管大学建筑类型学拒绝传统纪念碑性的标准，然而也没有能够找到与有形的材料和使用者无形的健康有关的技术解决方案，来创建自己特有的纪念碑性品牌。大学建筑需要适当的照明、能源节约、不受建筑结构影响的宽敞的空间以及与城市环境之间的渗透性。在最好的建筑设计案例中，这些要求在复杂而多变的建筑设计中都得到了满足。

大学建筑项目设计的关键理念包括整合和宜居。整个大学这一社区中的教师、学生、研究人员和行政人员都要能够充分体验和享受大学复杂的职能和作用，而这些人的要求又各不相同，有的要求为从事个人研究提供隐秘的空间，有的要求能提供可以聚集很多人的场所，有的要求技术实验室的高效运作，还有的要求能够获得可打印的和数字化的文档。然而，大学不仅仅是一处工作的地方，所以宜居这一设计理念从纯粹的技术特征层面延伸扩展到提供促进人们文化和社会交往的地方这一定义。

因此，供人们放松和进餐的空间在大学建筑中就扮演了主要角色。当代知识中大部分新思想都是在与人交往的过程中产生的。而在大学中，这些沟通交流不仅在实验室和学术空间里发生，也在大学的餐厅内发生。在很多更复杂的设计案例中，功能的整合和宜居绝不仅仅体现在学生公寓和运动、娱乐设施方面。

正如其他类型的当代建筑，大学建筑也特别强调与城市区域、大学社区等周边环境之间的渗透性。有时候，大学位于人口或多或少的乡村，也要注意其与所在地之间的渗透性。传统的建筑设计理念是设计一个人们可以进出、光线很好的容器。与此相比，大学建筑将自身分解，根据需要或凹进或掏空，实现与室外空间的连续统一。例如，如果学校的许多学术性空间和社交空间都俯瞰一条公路，那么大学建筑设计中就会考虑如何与这条公路取得空间上的连续性。

安藤忠雄在墨西哥蒙特雷大学RGS中心的设计中，虽没有牺牲严格的建筑构造特性，却实现了上述追求。RGS中心这栋看起来非常中规中矩的混凝土建筑内部却存在着无形的力量，将其余的建筑体量掏空、扭曲。这样，这个僵硬死板的几何体向外敞开自己，形成大型上空体量，但并不是简简单单的开口，而是非常开放的空间，其大小与建筑内的功能并不成比例。这一建筑的入口实现了与外部空间的无缝连接。与此同时，这一掏空的建筑体量不仅极大地凸显了这一巨大混凝土建筑的结构张力，内部的实验室和教室也可俯瞰由此形成的中央空间，因此建筑内部和外部之间的界定变得更加模糊。这种干预，超出了传统意义上的建筑，在完全实现建筑的可识别性和简洁紧致性的同时，成为蒙特雷大学校园道路交通系统中一道亮丽的风景。另外，如同整个蒙特雷大学的其他建筑，RGS中心也没有根据正交矩阵来确定朝向。安藤的建筑设计语言的表达力充分体现了当代纪念碑性的特性，赋予建筑的绝不仅仅是单纯的功能性，而是使建筑成为大学学府的地标，用无形的语言将其颂扬。

由Taller Básico de Arquitectura设计的圣豪尔赫大学卫生学院项目更加明确地强调建筑与周围环境的关系。圣豪尔赫大学卫生学院位于西班牙萨拉戈萨市附近的维拉纽瓦·德·加利西亚小镇的边缘，一

volumes. In this way the rigid prismatic geometry opens itself outward, creating large voids that are not simple openings, but open spaces whose size is out of scale with respect to the functions hosted, an invitation to the entrance in seamless relation with the outside. At the same time, the emptying of the volumes not only dramatically highlights the structural tensions of the large block of concrete, but allows laboratories and classrooms to overlook the central space, thus accentuating even more ambiguity between inside and outside. The intervention, while fully maintaining its recognizability and compactness, becomes more an episode of the complex Monterrey campus system of paths than a building in the traditional sense of the term. This feeling is enhanced by the refusal to orient according to an orthogonal matrix, a refusal shared with other buildings of the university compound. The expressive power of Ando's language achieves a character of contemporary monumentality that gives to the building more than mere functionality, making it a landmark that celebrates the institution of the university with no rhetoric.

The program of Taller Básico de Arquitectura for the Health Faculty of San Jorge University is even more explicitly oriented to the relationship with the surrounding environment. Located at the edge of the small town of Villanueva de Gallego, near Zaragoza, Spain, the faculty faces on one side an industrial area and on the other open countryside. Although not exploiting the amazing view of the Pacific Ocean of which Louis Kahn was able to take advantage in designing the Salk Laboratories, the authors have nevertheless opted for a facade solution that imposes precise visual directions on those who observe from inside of the building, while partializing the direct sunlight. The scaled sequence of white inclined walls that define those directions presents itself as a fragmented surface that does not impose itself on the bare landscape but instead becomes a fundamental element of a "new naturality". The overall layout generates, through three concave bodies, non-orthogonal or parallel, a sort of minimalist urban nucleus, wherein the solid volumes lead to a "square", central space for academic community relationships that overlooks classroom corridors. The functional articulation is extremely simple, offering no specialized spaces for classrooms and laboratories, but rather providing sizes and shapes that offer flexibility for future reorganization, while creating spaces for relational activities, such as the cafeteria and conference room, in the creases of the L-shaped buildings.

Although the University of Limerick, Ireland, was relatively speaking founded quite recently, its layout belongs to the Anglo-Saxon tradition of complex and articulated campuses that include buildings for teaching and research, student residences, social and welfare services and sports and recreational facilities, all located in a pleasant landscape of meadows and oak trees. The complex, designed by Grafton Architects, includes the Medical School and

1. Leslie T., *Louis I. Kahn: Building Art, Building Science*, New York, 2005.
2. Krunic D., *The Groundscraper: Candilis-Josic-Woods Free University Building, Berlin 1963-1973*, Los Angeles: University of California, 2012.
3. Jacobus J., "Engineering Building, Leicester University", Architectural Review, Vol.806, 1964.

边面临的是工业区，另一边是广阔的乡村。虽然无法像路易斯·康那样借助太平洋的旖旎风光来设计索尔克研究所的实验室，不过设计师们选择的外立面解决方案不仅使从建筑内向外看的人们获得精确的视觉方向，同时也使室内避免了阳光的直射。定义视觉方向的那一面白色的犹如鱼鳞状的斜墙，其碎片状的表面在周围裸露的景观中并不显得突兀，相反却成为"新自然性"的基本元素。通过三个向内凹进、既非正交也非平行的建筑体量，卫生学院的整体布局形成一种带有极简主义的城市核心，三个建筑体量围成了一个"广场"，成为学校中人与人交往的中央空间，教室的走廊都面朝着这一中央空间。功能空间设计得非常简单，没有专门用作教室和实验室的空间设计，而是提供不同尺寸与形状的房间，将来可根据需要灵活安排，同时设计了许多人际交往的空间，如在L形建筑的接合部，设计有餐厅和会议室。

尽管爱尔兰利默里克大学相对来说成立较晚，但其布局属于传统的盎格鲁—撒克逊校园风格，布局复杂而清晰，包括教学楼和研究实验楼、学生公寓、社会和福利服务设施、运动和娱乐设施，所有这一切都位于令人心旷神怡的充满草坪和橡树的校园中。由格拉夫顿建筑事务所设计的医学院及其学生公寓位于这样一个强大的功能整合的校园中。其建筑布局符合校园功能整合的"城市"特性，包括一系列自由布局的建筑，人行步道和大型公共开放空间将各座建筑连为一体。大型的公共开放空间为人们提供了超越纯学术交流活动使用的社会生活场所。此设计中还包括一个公共汽车站和自行车存放处。其形状与校园中建筑的自由布局风格完全一致。医学院及其学生公寓邻近运动馆、爱尔兰世界音乐和舞蹈学院以及健康科学学院楼，进一步体现了其功能整合的设计理念。医学院和学生公寓的外表面都斜切凹进，形成厚厚的三维洞口，在视觉上要比普通的镶嵌在建筑外表的二维窗户更能达到与周围公共空间保持连贯、融为一体的效果。屋顶的形状可以设计出大型天窗，为室内公共空间提供光源。为了突出不同建筑的目标和功能，医学院和学生公寓外立面的处理和材料的选择也是有区别的。

在CRAB工作室设计的维也纳大学法学院和中央行政大楼的项目中，显而易见的是，当代教育活动的理念远远超出了简单的课堂教学。大学生活的每一刻钟都是学习经历的一部分，将教室连为一体的公共空间同样重要，教师们和学生们彼此之间可以在此见面，相互交流。为了突出整个空间的活力，CRAB工作室集约利用颜色和定制家具来消除不同领域空间之间的层次结构。多层平面和内部体量那富有曲线的、有机的、非正交的形态结构进一步加强了建筑的社交性和功能性的整合。学校活动延伸到室外空间，即延伸到阳台和屋顶露台。室外空间并不局限于一楼地面。通过坡道系统，室外空间将城市景观与建筑的多个层面连为一体，成为建筑不可分割的组成部分。五颜六色的外墙立面在方向不一的木质百叶的保护下，免受阳光的直射。随着时间的推移，建筑表面将被自然植被覆盖，成为周围环境的一部分。

从上述这些例子以及从最近设计的大学和研究机构的建筑的一般情况来看，很显然，传播知识的机构不再代表神圣和权力，而是明确主张大学建筑应该成为教师、研究人员和学生高度参与的社区体系的一部分，这并没有忽视超出大学社区之外的强大整合。

its Student Residences in the logic of a strong functional integration. Its layout is consistent with the integrated "urban" character of the campus, consisting of a series of freely arranged buildings connected by pedestrian paths and large public open spaces that encourage a social life that goes beyond pure academic activity. Also provided is a Bus and Bicycle Shelter whose shape is in accord with the free arrangement of the buildings on the site. The concept of integration is enhanced by close proximity to the Sports Pavilion, the Irish World Academy of Music and Dance and the Health Sciences Building. Both the Medical School and the residential buildings are treated as volumes whose outer surfaces are carved by oblique cuts in the thickness of the openings that give a three-dimensional appearance, connecting them in a much stronger way with the surrounding public space than would have been possible with regular two-dimensional windows inserted in the outer surface. The shape of the roof permits the use of large skylights for lighting the interior collective spaces. The treatment of the facades and their material choices are differentiated between the academic building and the residences to highlight the destination and the functional complexity of the compound.
In the project for Vienna University's Law and Administration Buildings by CRAB Studio, it is evident that the contemporary concept of educational activities extends far beyond the simple frontal teaching in a classroom. Every moment in the Faculty's life becomes part of the learning experience and the common spaces connecting the classrooms play an equally important role for meeting and mutual exchanges among teachers and students. To accentuate the dynamism of the entire spatial system, CRAB Studio makes intensive use of colors and bespoke furniture to eliminate any hierarchy among the various areas. The social and functional integration of the buildings is enhanced by the curvaceous, organic and non-orthogonal morphology of the multi-level floor plans and of the internal volumes. The activities extend to the outdoor spaces, as balconies and roof terraces. The outdoor spaces are not limited to the ground floor but, through a system of ramps, they connect the urban landscape to the multiple levels of the buildings of which become an integral part. The colorful facades are protected from the sun by a multi-oriented system of timber louvers and over time the structure will be covered in natural vegetation, allowing the design to become part of the surrounding environment.
From these examples, and from the general scenario of recent architecture for academic and research institutions, it is apparent that the abandonment of a sacred, authoritarian representation of knowledge institutions and the definitive affirmation of a design conceived as part of community systems for high participation of teachers, researchers and students do not neglect a strong integration that extends to the extra-academic community. Aldo Vanini

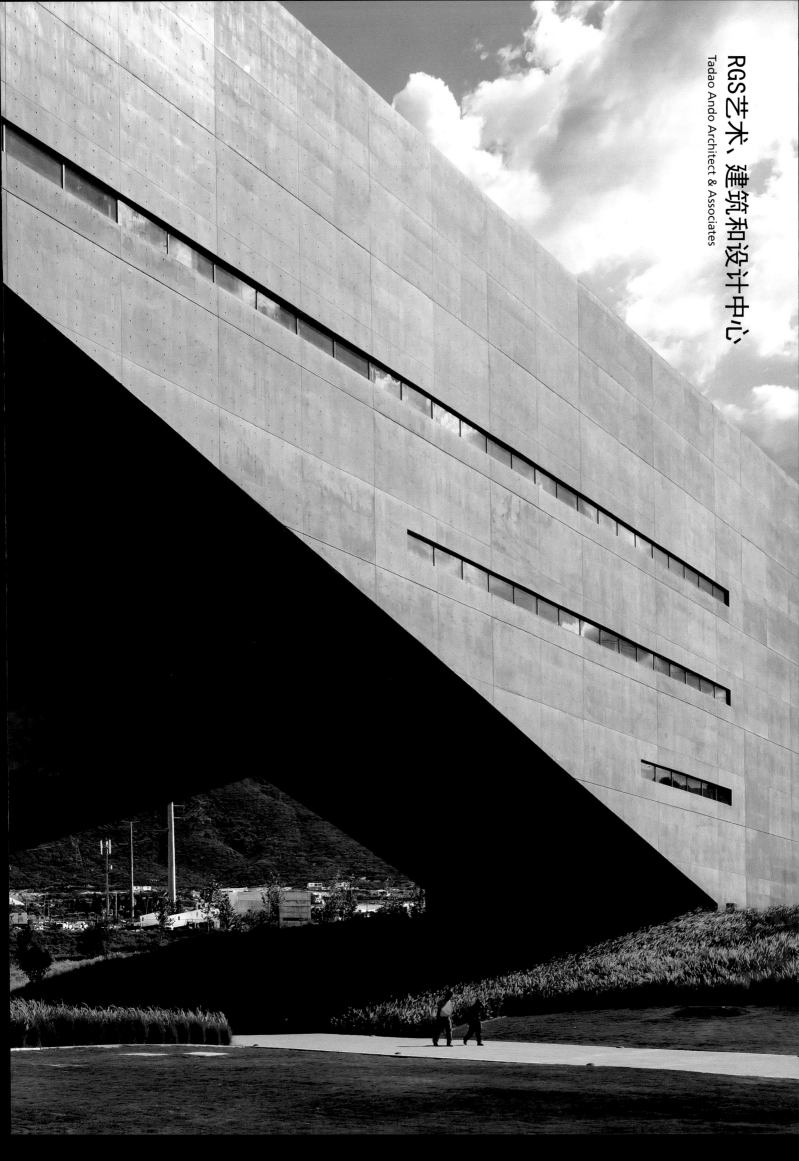

# RGS艺术、建筑和设计中心
Tadao Ando Architect & Associates

墨西哥蒙特雷大学位于其第二大工业城市,是一所拥有12 000名学生的罗马天主教大学。RGS中心是为新成立的艺术、建筑和设计学院所设计。

项目所在地很显眼,位于学校北面,靠近学校入口处。建筑师的设计想法是建造一座通道形状的建筑,既欢迎八方来客,又可饱览蒙特雷高山国家公园的旖旎风光。

这座99m×27m×32.4m的长方体建筑,通过两个透视的外壳形成其体量,这两个建筑外壳是利用旋转沿长轴设置的对角线来获得的。设计师安藤忠雄早在阿布扎比海事博物馆(2006年至现在)的设计中就使用过这一设计理念,不过其双曲线是沿短轴旋转。然而,在RGS中心设计中,双曲线是沿长轴旋转的,因此,其形成的外壳造型更加复杂,更加富有活力。室内空间根据所构成的复杂剖面外形而进行布局,这些外形通过贯穿整座建筑的公共空间,在三维空间内交织缠绕为一体。有些公共空间面向室外空间开放,使建筑内充满自然光和风。

此建筑被命名为"创意之门",希望为年轻人敞开大门,把他们送往人生的下一站。

## Roberto Garza Sada Center for Arts, Architecture and Design

University of Monterrey located in the second largest industrial city in Mexico is a Roman Catholic institution with 12,000 students. The RGS Center is a building which was designed for newly-established schools of art, architecture and design.

The project site is in a characteristic location in the northern part of the campus near the university entrance. The idea the architects proposed was to build a gateway-shaped architecture which will welcome people and command the spectacular view of the mountains of Parque Nacional Cumbres de Monterrey.

The form of the building is a 99m x 27m x 32.4m rectangular block with the void defined by a pair of hyper shells obtained by rotating diagonal lines along the longer axis. With the same concept, the form of the Abu Dhabi Maritime Museum (2006-) was created by a pair of hyper shells rotated along the shorter axis. However, in this project, the rotation is along the longer axis, resulting in the more complex and dynamic shell form. The interior space is organized by taking advantage of complex sectional shapes, and they are three-dimensionally intertwining with each other through public spaces which cut across the building. Some of those public spaces are open to exterior space, introducing natural light and wind to the building.

The building is named "Gate of Creation", hoping to welcome young people and to send them off to the next step of their lives.

Tadao Ando Architect & Associates

项目名称:Roberto Garza Sada Center for Arts, Architecture and Design
地点:Monterrey, Mexico
建筑师:Tadao Ando, Kazuya Okano
登记建筑师:Alexandre Le Noir, Vicente Tapia
项目管理:Raul Berarducci, Patricio Coen Mitrani
结构工程师:WSP Cantor Seinuk
机械工程师:IPISA
电气工程师:IEMMSA
HVAC工程师:PROYECTAIRE
照明顾问:Illumination Total
景观建筑师:SWA
LEED认证:Spiezle Group
总承包商:Constructora Garza Ponce
电气安装:IOESA
消防:GINSATEC
网络:DAS integration
混凝土供应商:CEMEX
甲方:Universidad de Monterrey
用地面积:20,700m²
有效楼层面积:12,693.52m²
设计时间:2007.4—2009.7
施工时间:2009.7—2012.12
摄影师:©Shigeo Ogawa

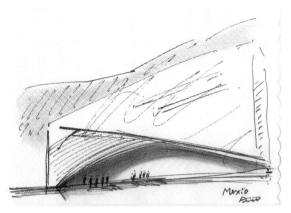

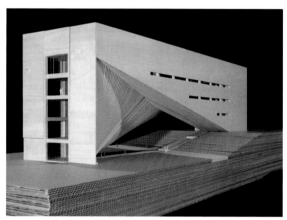

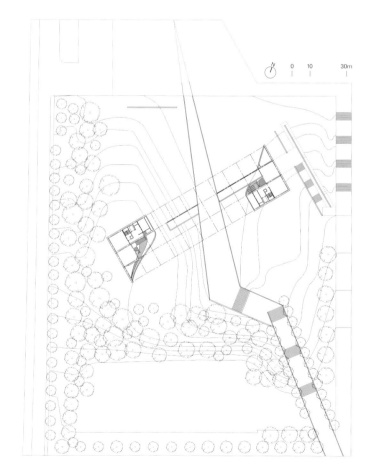

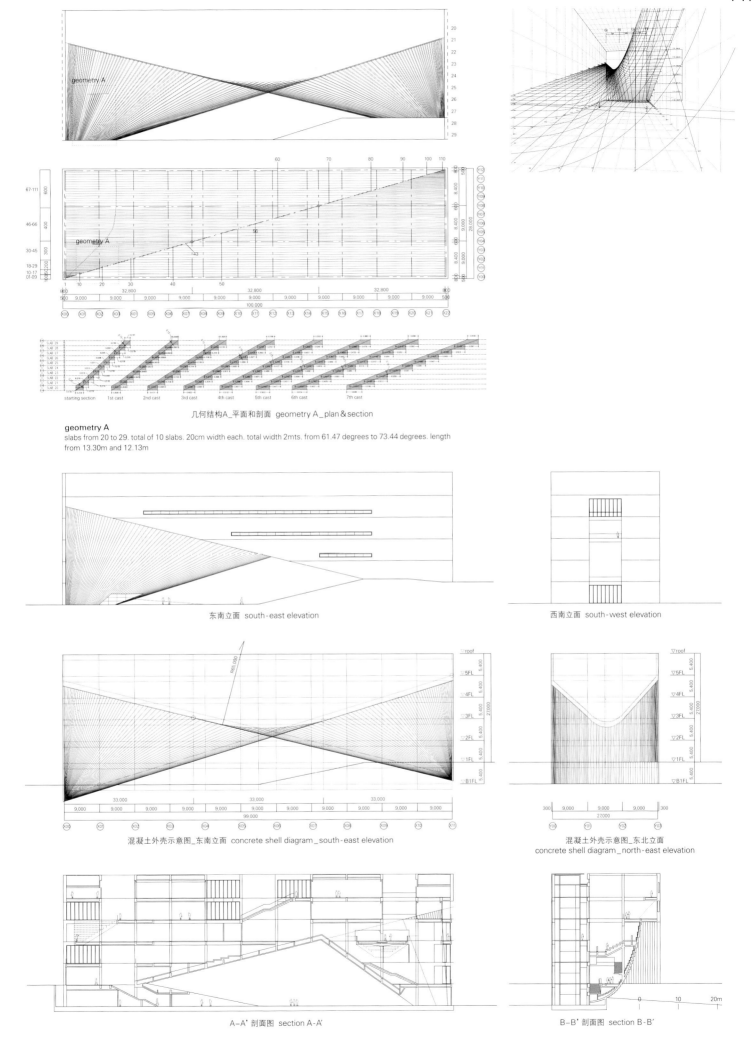

几何结构A_平面和剖面 geometry A_plan & section

**geometry A**
slabs from 20 to 29. total of 10 slabs. 20cm width each. total width 2mts. from 61.47 degrees to 73.44 degrees. length from 13.30m and 12.13m

东南立面 south-east elevation

西南立面 south-west elevation

混凝土外壳示意图_东南立面 concrete shell diagram_south-east elevation

混凝土外壳示意图_东北立面 concrete shell diagram_north-east elevation

A-A' 剖面图 section A-A'

B-B' 剖面图 section B-B'

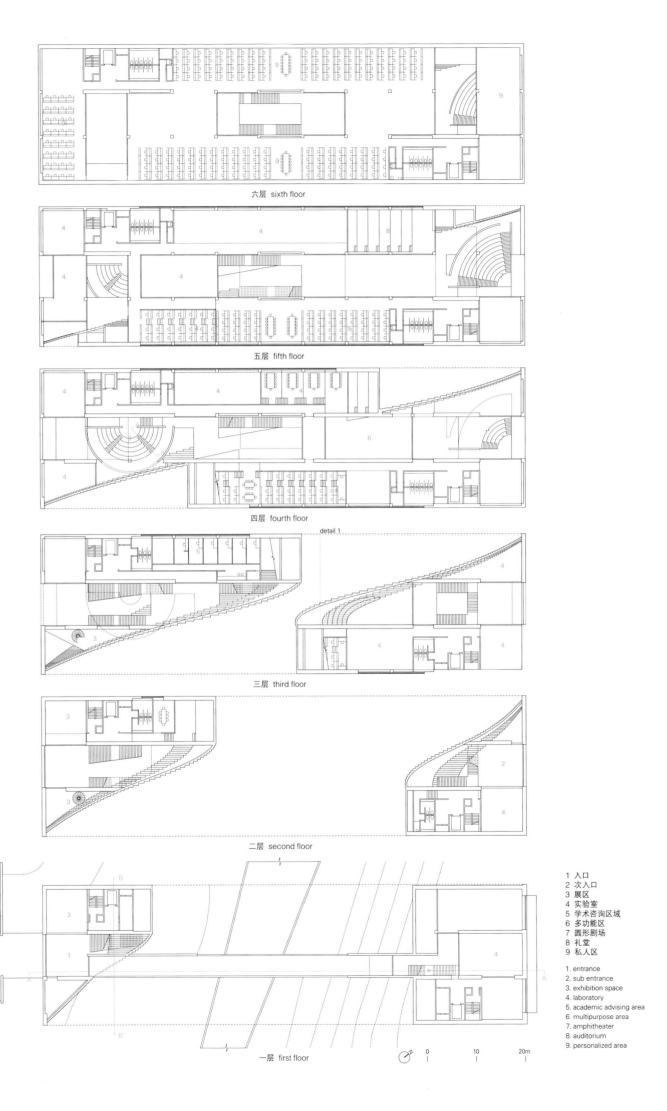

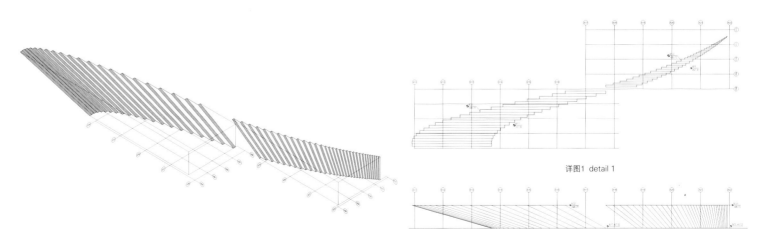

详图1 detail 1

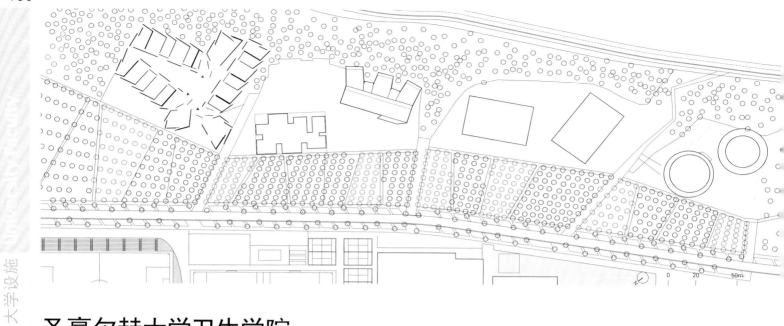

# 圣豪尔赫大学卫生学院
Taller Básico de Arquitectura

新建的圣豪尔赫大学卫生学院坐落在萨拉戈萨市的郊区。虽然校园位于郊区，但自然的景色资源稀缺。沿着校园的郁郁葱葱的森林也是人工修建而成的。周围的建筑物还有礼堂和通信学院，与位于森林一侧、同时期建造的一座建筑相互应和。

卫生学院加入并拓展了现有建筑物所在的整个自然区域空间，将来还会有新的建筑加入。这一新学院不仅仅是一座大楼，更是成为这一全新地方的一部分。而这幢建筑也被视为新自然景观的一部分。

建筑设计方案以三条凹线布置展开。这些白色的如鱼鳞状的线条在校园内展开，成为校园景观的一部分。共有二层的建筑内部精心布置了用于教学和科研的教室和实验室。鱼鳞外形为每个房间提供了所需的自然光线。房间的尺寸和形状为其灵活使用提供了百变空间，因此可以很方便地根据教学和科研要求来重新布局。通过鱼鳞外形进入室内的光线是可控的，房间内可以使用各种数字技术而不受光的影响。线形折叠处是新学院的公共空间所在，有餐厅、会议室和多功能室。

三条线围成一处向天空敞开的大空间。所有进入实验室和教室的走廊都面对着这处开阔的空间。穿过广场可以到达三条线的位置。每一条线通过广场相互照应，广场构成整个学院建筑群的内部。建筑的内外关系是反向的，外面的凹陷空间恰好是室内空间，而里面的凸出空间则变成最外部的地方。

圣豪尔赫大学卫生学院外部白色鱼鳞状的新型景观既吸纳了自然光线，又成为大学一道亮丽的风景，同时该建筑群内部也提供了一处向天空敞开的大空间。

## Health Faculty of San Jorge University

The new Health Faculty of San Jorge University is located on a campus on the outskirts of Zaragoza City. Although it is a rural campus, the nature in it is scarce. The forest along the campus is the result of a man-created operation. The surrounding buildings, the Rectory and Communications Faculty, respond to a contemporaneous architecture that lives besides that nature.

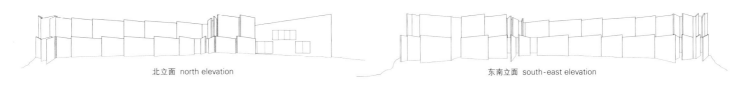

北立面 north elevation　　　　　　东南立面 south-east elevation

项目名称：Health Sciences Faculty
地点：San Jorge University Campus, Highway A-23 Zaragoza-Huesca KM 299, CP 50830, Villanueva de Gallego
建筑师：Javier Pérez-Herreras, Javier Quintana de Uña
合作建筑师：Edurne Pérez Daz de Arcaya, David Santamaria Ozcoidi, Leire Zaldua Amundarain, Xabier Ilundain Madurga, Daniel Ruiz de Gordejuela Telleche
结构工程师：FS Estructuras  系统工程师：GE&Asociados
景观建筑师：Taller Básico de Arquitectura
开发商：Universidad San Jorge Foundation  施工单位：San Jorge UTE
面积：8,853m²  施工时间：2010.2—2012.6
摄影师：©José Manuel Cutillas (courtesy of the architect)

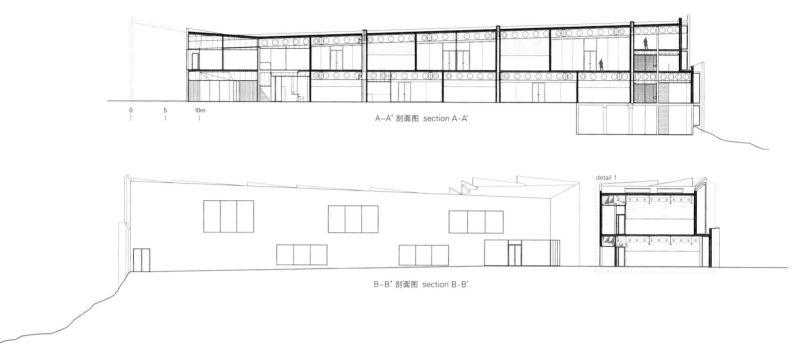

C-C' 剖面图 section C-C'

D-D' 剖面图 section D-D'

The Health Faculty joins the development of that little nature to reinforce the place where the existing buildings rest and where new buildings will do. The new faculty is not only another building; it becomes part of the new place. Architecture is thought as part of a new nature.

The building program is organized in three concave lines. These white and scaled lines unfold on the campus as part of its landscape. Inside, on two floors, classrooms and laboratories are organized for teaching and research. Each scale catches the light needed for each room. The dimensions and shape of rooms allow a big variability of use. Consequently, it is possible for an academic reorganization in an easy way. Light coming through scales can be controlled, so digital technologies can be used inside rooms. The creases of each line contain the most public rooms of the new faculty: cafeteria, conference room and multipurpose rooms.

The three lines enclose a big room open to the sky. All the access corridors to laboratories and classrooms face this big room. The square gives access to the three lines. Lines look at each other through the square, which discovers the inside of this complex. The inside and outside relation of the faculty gets inverted. The concave outside happens to be the most interior room, and the convex inside becomes the most exterior place.

The mineral nature of this faculty in San Jorge University offers a new landscape of white scales breathing light on the outside, and it offers a big room opened to the sky on the inside.

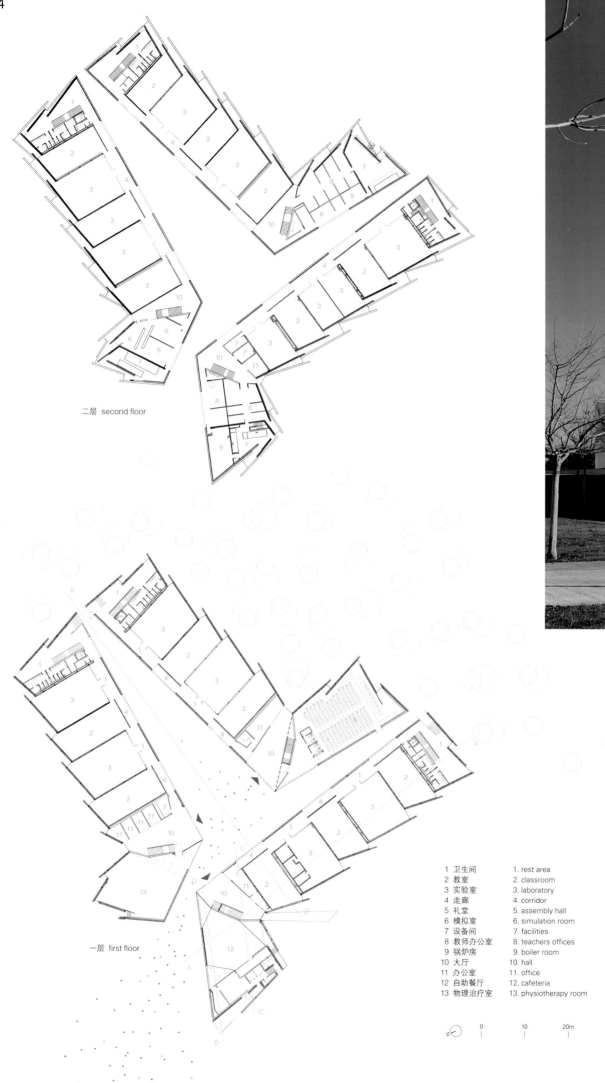

二层 second floor

一层 first floor

| | | |
|---|---|---|
| 1 | 卫生间 | 1. rest area |
| 2 | 教室 | 2. classroom |
| 3 | 实验室 | 3. laboratory |
| 4 | 走廊 | 4. corridor |
| 5 | 礼堂 | 5. assembly hall |
| 6 | 模拟室 | 6. simulation room |
| 7 | 设备间 | 7. facilities |
| 8 | 教师办公室 | 8. teachers offices |
| 9 | 锅炉房 | 9. boiler room |
| 10 | 大厅 | 10. hall |
| 11 | 办公室 | 11. office |
| 12 | 自助餐厅 | 12. cafeteria |
| 13 | 物理治疗室 | 13. physiotherapy room |

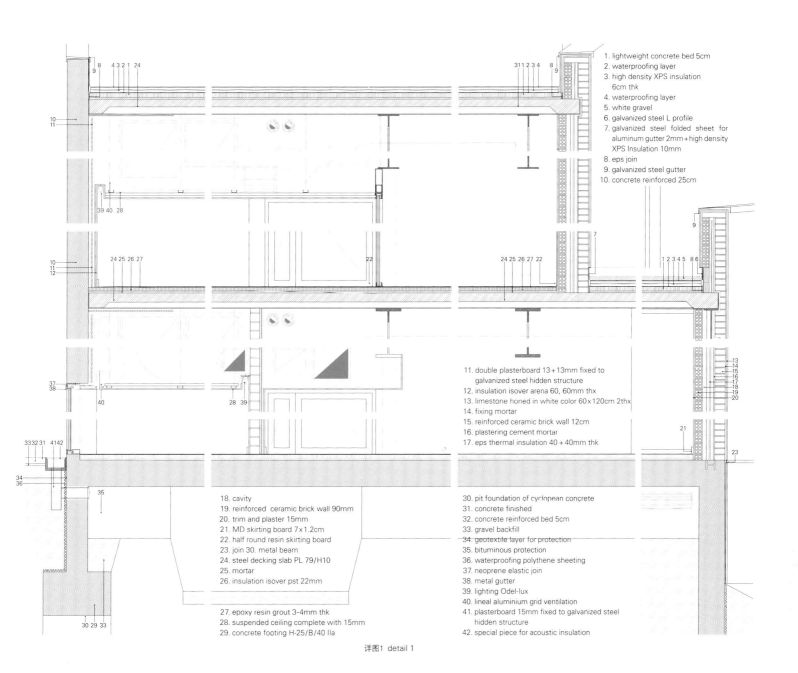

1. lightweight concrete bed 5cm
2. waterproofing layer
3. high density XPS insulation 6cm thk
4. waterproofing layer
5. white gravel
6. galvanized steel L profile
7. galvanized steel folded sheet for aluminum gutter 2mm + high density XPS Insulation 10mm
8. eps join
9. galvanized steel gutter
10. concrete reinforced 25cm
11. double plasterboard 13+13mm fixed to galvanized steel hidden structure
12. insulation isover arena 60, 60mm thx
13. limestone honed in white color 60x120cm 2thx
14. fixing mortar
15. reinforced ceramic brick wall 12cm
16. plastering cement mortar
17. eps thermal insulation 40 + 40mm thk
18. cavity
19. reinforced ceramic brick wall 90mm
20. trim and plaster 15mm
21. MD skirting board 7x1.2cm
22. half round resin skirting board
23. join 30. metal beam
24. steel decking slab PL 79/H10
25. mortar
26. insulation isover pst 22mm
27. epoxy resin grout 3-4mm thk
28. suspended ceiling complete with 15mm
29. concrete footing H-25/B/40 IIa
30. pit foundation of cyclopean concrete
31. concrete finished
32. concrete reinforced bed 5cm
33. gravel backfill
34. geotextile layer for protection
35. bituminous protection
36. waterproofing polythene sheeting
37. neoprene elastic join
38. metal gutter
39. lighting Odel-lux
40. lineal aluminium grid ventilation
41. plasterboard 15mm fixed to galvanized steel hidden structure
42. special piece for acoustic insulation

详图1 detail 1

# 利默里克大学医学院和学生公寓

Grafton Architects

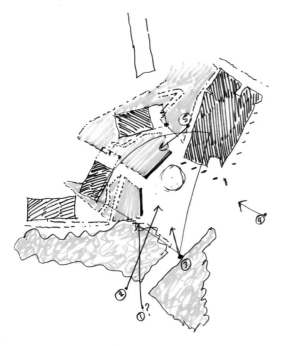

利默里克大学位于爱尔兰西南部,校园占地面积很大,以前是一处私人领地,爱尔兰最大、最长的河流香农河的下游穿过校园。校园最近往河的北面拓展扩建,新建一栋医学院大楼和医学院学生宿舍楼,通过人行天桥与原校区相连。这些新建筑的设计也意在营造一处大型公共开放空间,使其最终成为香农河北部扩建校区的焦点。

设计理念是将医学院大楼与学生宿舍楼融为一体,相互重叠,致力于大学里公共生活空间的发展。

此项目的整体特点基于校园的总体规划要求,即在香农河北岸使用有机方法来构建公共生活空间。而场地地势倾斜,原有农业耕地景观的残余模式依然清晰可见,有大量农田和灌木篱墙。

这一新建筑群将与附近三座原有的学院楼建筑,即运动馆、爱尔兰世界音乐和舞蹈学院以及健康科学学院楼一起,共同营造一处新的公共空间。

新建筑群由医学院大楼、三栋学生宿舍楼和用来存放自行车和遮盖公交车的天篷/绿廊组成。

医学院大楼是校园一系列规划设计中的最后一个版块,相当于一个锚点,现在其他建筑物都松散地围绕在其四周。医学院大楼的设计语言就是一个教育机构所应用的语言,而学生宿舍楼看起来像三座大房子。混凝土公交车车篷、学生宿舍楼和医学院大楼松散而有序地围绕公共空间而建。公交车车篷、台阶和坡道使这一公共空间与远处的运动馆呈现平缓的过渡。

中央空间东高西低,坡度平缓。三棵橡树、石头坐椅和台阶占据着中央平台的显著位置,在边缘处向四周发散开来,与学生宿舍楼、停车场和其他院系建筑的边界连为一体。公共空间的地面从硬到软,南面是坡度平缓的草坪,时有时无地设置一些设施,供人们休闲和散步徜徉。

各栋建筑如卫兵般面朝开放的公共空间站立,所用的建筑材料彰显彼此的不同。

石灰石是医学院大楼使用的主要建筑材料,主要是因为校园所在地克莱尔县境内盛产石灰石。石质外墙根据朝向、太阳、风、雨和公共活动的不同情况而折叠、变换形状和叠加。西南向的柱廊既是入口,又是公众聚会的空间。相比之下,东面和北面的墙稍显沉寂单调。

考虑到医学院大楼的整体平面设计,其楼顶设计进行了调整,来为多个空间提供自然光源,包括中央流线空间、临床实践技能实验室、走廊和一个小屋顶露台。

开放的中央楼梯既连接所有的主要空间,也将楼内各层的内部空间连为一体,它可以作为社交空间,大家在此停下脚步来聊天,或倚靠在栏杆上,看入口处的人来人往,或观看上面和下面的其他活动空间。

原有的宿舍楼都是砖砌的,新建的学生宿舍楼也与其保持一致。但新宿舍楼的砖的应用增加了厚度,外墙立面的窗户深深凹进,在室内和可以俯瞰的公共空间之间形成类似门槛的空间。所有学生生活空间中的公共空间都是东南向,而稍私密的卧室都是东北向或西北向。

宿舍楼的地下室设计成拱形门廊,为学生们提供了一处可遮风挡雨的社交空间,行人如果到校园北面也可以从停车场和公共汽车站台穿过。进入宿舍楼的通道很宽敞,并且通向宿舍楼的入户门厅,宿舍楼里设有楼梯、电梯、自行车存放处和公共洗衣设施。

### Medical School, Student Residences at the University of Limerick

The University of Limerick, in the Southwest of Ireland occupies a large territory, formerly a Demesne, and is situated on both sides of the lower reaches of the River Shannon, the longest and largest river in Ireland. Part of its most recent expansion to the north of this great river, accessible by pedestrian bridge from the existing

campus, provides for the construction of a new medical school building and accommodation buildings for students attending the facility. These new buildings are also intended to address a large public open space which will ultimately become the focal point for this expansion of the campus to the North.

The aspiration is to combine faculty buildings and residences in a manner which encourages overlap and contributes to the life of the public spaces at the University.

Aspects of the formal character are derived from an interpretation of the campus' masterplan which requires an organic approach to the making of public spaces on the north side of the River Shannon. Here the ground is sloping and remnants of the agrarian

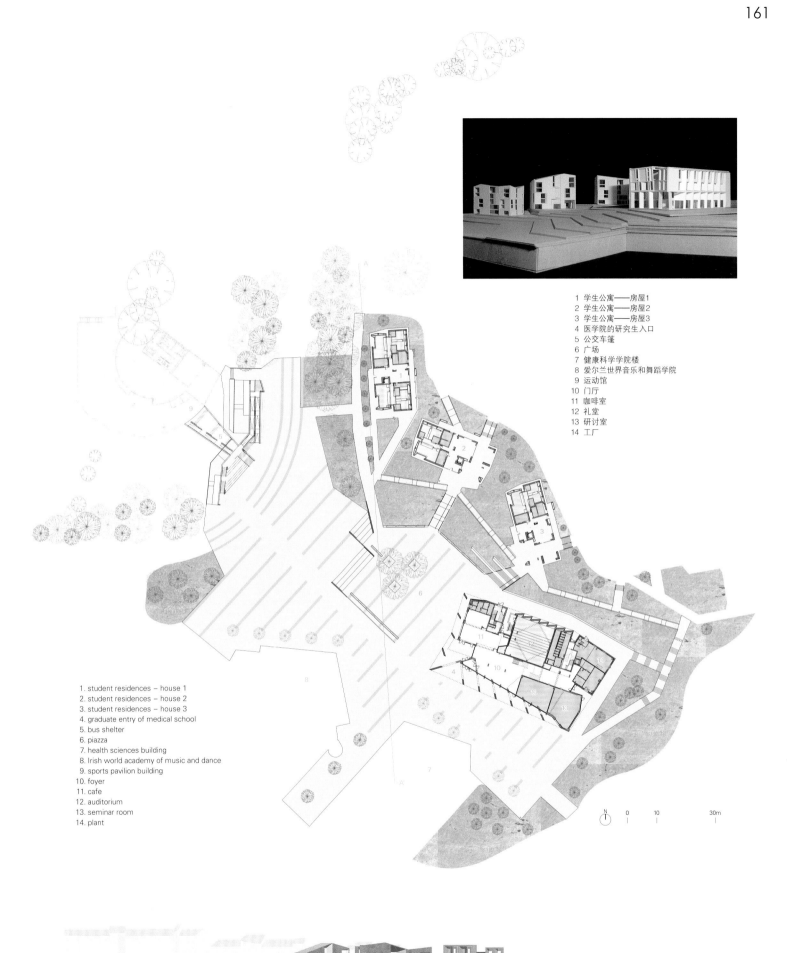

1. student residences – house 1
2. student residences – house 2
3. student residences – house 3
4. graduate entry of medical school
5. bus shelter
6. piazza
7. health sciences building
8. Irish world academy of music and dance
9. sports pavilion building
10. foyer
11. cafe
12. auditorium
13. seminar room
14. plant

A-A' 剖面图 section A-A'

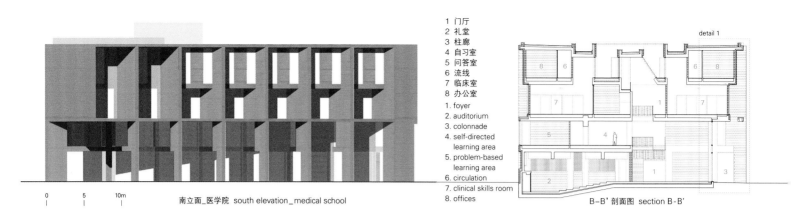

| | |
|---|---|
| 1 门厅 | 1. foyer |
| 2 礼堂 | 2. auditorium |
| 3 柱廊 | 3. colonnade |
| 4 自习室 | 4. self-directed learning area |
| 5 问答室 | 5. problem-based learning area |
| 6 流线 | 6. circulation |
| 7 临床室 | 7. clinical skills room |
| 8 办公室 | 8. offices |

南立面_医学院 south elevation_medical school

B-B' 剖面图 section B-B'

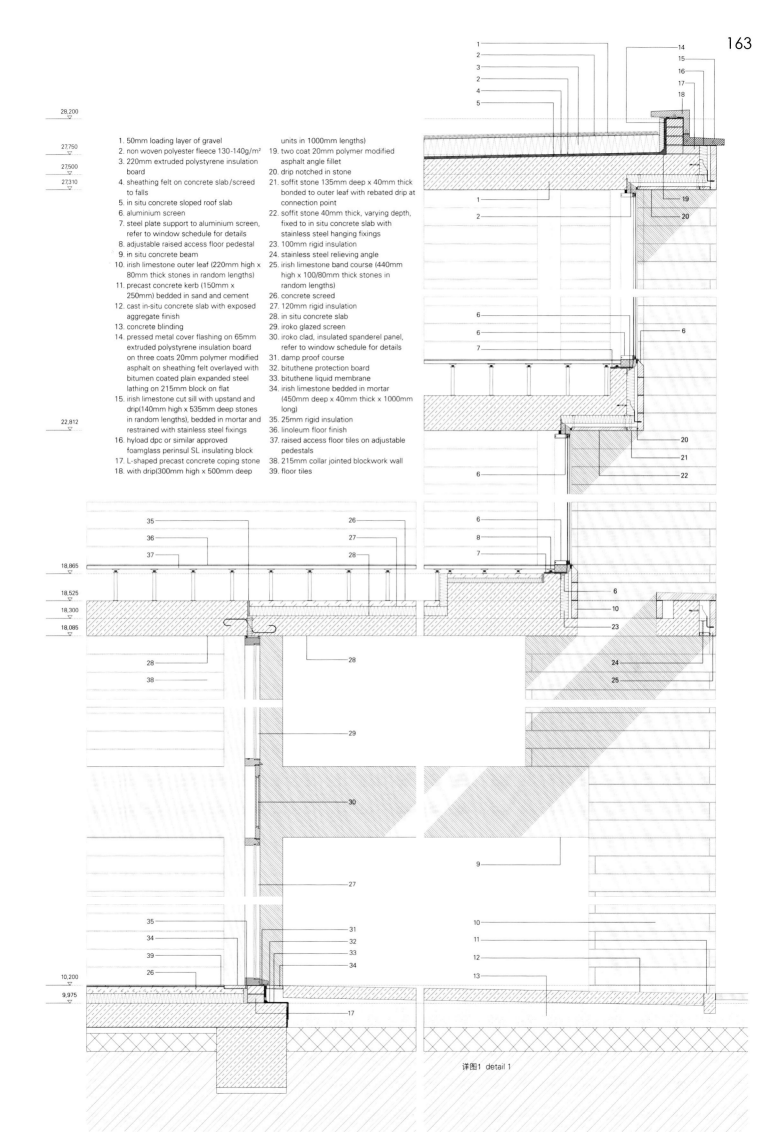

1. 50mm loading layer of gravel
2. non woven polyester fleece 130-140g/m²
3. 220mm extruded polystyrene insulation board
4. sheathing felt on concrete slab/screed to falls
5. in situ concrete sloped roof slab
6. aluminium screen
7. steel plate support to aluminium screen, refer to window schedule for details
8. adjustable raised access floor pedestal
9. in situ concrete beam
10. irish limestone outer leaf (220mm high x 80mm thick stones in random lengths)
11. precast concrete kerb (150mm x 250mm) bedded in sand and cement
12. cast in-situ concrete slab with exposed aggregate finish
13. concrete blinding
14. pressed metal cover flashing on 65mm extruded polystyrene insulation board on three coats 20mm polymer modified asphalt on sheathing felt overlayed with bitumen coated plain expanded steel lathing on 215mm block on flat
15. irish limestone cut sill with upstand and drip(140mm high x 535mm deep stones in random lengths), bedded in mortar and restrained with stainless steel fixings
16. hyload dpc or similar approved foamglass perinsul SL insulating block
17. L-shaped precast concrete coping stone
18. with drip(300mm high x 500mm deep units in 1000mm lengths)
19. two coat 20mm polymer modified asphalt angle fillet
20. drip notched in stone
21. soffit stone 135mm deep x 40mm thick bonded to outer leaf with rebated drip at connection point
22. soffit stone 40mm thick, varying depth, fixed to in situ concrete slab with stainless steel hanging fixings
23. 100mm rigid insulation
24. stainless steel relieving angle
25. irish limestone band course (440mm high x 100/80mm thick stones in random lengths)
26. concrete screed
27. 120mm rigid insulation
28. in situ concrete slab
29. iroko glazed screen
30. iroko clad, insulated spanderel panel, refer to window schedule for details
31. damp proof course
32. bituthene protection board
33. bituthene liquid membrane
34. irish limestone bedded in mortar (450mm deep x 40mm thick x 1000mm long)
35. 25mm rigid insulation
36. linoleum floor finish
37. raised access floor tiles on adjustable pedestals
38. 215mm collar jointed blockwork wall
39. floor tiles

详图1 detail 1

landscape pattern are still evident in the form of old field patterns and hedgerows.

This new suite of buildings combines with three existing, neighboring institutions, the Sports Pavilion, the Irish World Academy of Music and Dance and the Health Sciences Building, in order to make a new public space.

The new buildings consist of a medical school, three blocks of student housing and a canopy/pergola forming a bus and bicycle shelter.

The Medical School, the last in a series of set pieces, acts as an anchor around which the other buildings now loosely rotate. The language of the medical school is that of an educational institution while the student residences appear like three large houses. The concrete bus shelter, together with the residences combines with the medical school to form a loose edge to the public space. The bus shelter canopy, steps and ramps negotiate the level change to the sports pavilion beyond.

The central space slopes gently to the west. Three oak trees, stone seats and steps occupy a central level platform subtly providing a focal point, fracturing at the edges to connect to the residences, car parking and other faculty buildings. The surfaces of the public space move from hard to soft; south sloping grassed spaces, designed with and without furniture, provide for leisure and lingering.

The buildings stand like guards facing the public space, distinguished by their materials.

Limestone is used to represent the "formal" central medical school, making reference to the limestone territory of County Clare in which this side of the campus is located. The stone wall is folded, profiled and layered in response to orientation, sun, wind, rain and public activity. A colonnade to the south and west corner acts as a gathering and entrance space. In contrast the north and east walls are more mute.

In response to the deep plan, the roof-form is modulated to light multiple spaces, including the central circulation space, the clinical skills labs, the corridors, and a small roof terrace.

An open central stair connecting all of the primary spaces, threads through all levels of the interior, is designed as a social space with enough rooms to stop and chat or lean on a balustrade/shelf and view the activity of the entrance and other spaces above and below.

Brick follows through to the residences from the existing accommodation buildings behind. Here the material is given depth and the facades are deeply carved providing a form of threshold between the domestic interior and the public space that they overlook. All living spaces address the public space to the southeast with the more private study bedrooms facing northeast or northwest.

The undercroft of the residences is carved away providing archways allowing pedestrian movement from the car park and bus park to the north as well as forming sheltered social spaces for students. Large gateways open into the entrance courts of the housing blocks where stairs, lift, bicycles bins and common laundry facilities are.

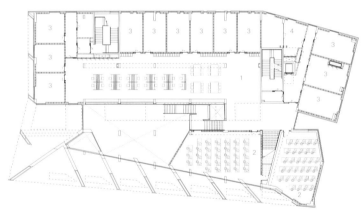

1 自习室  2 计算机室  3 问答室  4 全科医生中心
1. self-directed learning area  2. computer room  3. problem-base learning area  4. GP core
二层 second floor

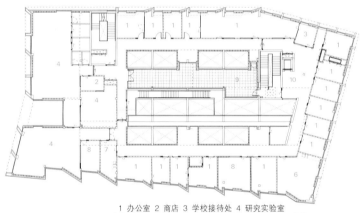

1 办公室  2 商店  3 学校接待处  4 研究实验室
5 研究报告区  6 员工室  7 厨房  8 会议室  9 庭院  10 流线
1. offices  2. store  3. school reception  4. research laboratory
5. research write-up area  6. staff room  7. kitchen  8. meeting room  9. courtyard  10. circulation
四层 fourth floor

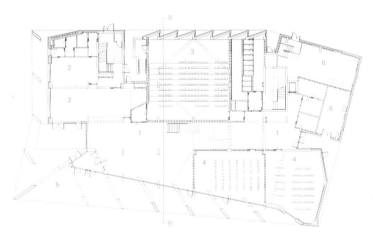

1 门厅  2 咖啡室  3 礼堂  4 研讨室  5 柱廊  6 商店/工厂
1. foyer  2. cafe  3. auditorium  4. seminar room  5. colonnade  6. store/plant
一层 first floor

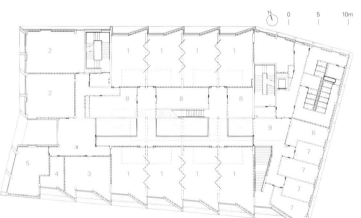

1 临床室  2 解剖室  3 计算机实验室
4 临床录像室  5 自习室  6 公用办公桌  7 办公室  8 商店  9 流线
1. clinical skills room  2. anatomical skills room  3. computer lab
4. clinical skills video room  5. self-directed learning area  6. hot desks  7. offices  8. store  9. circulation
三层 third floor

项目名称：Medical School, Student Residences and Bus Shelter at the University of Limerick
地点：University of Limerick, Limerick, Ireland
建筑师：Grafton Architects
项目团队：Yvonne Farrell, Ger Carty(project director), Philippe O'Sullivan
Matt McCullagh(project architect of student housing, piazza, pergola/ bus shelter)
Kieran O'Brien(project architect of medical school)
Abi Hudson(project architect of medical school)
David Healy(assistant project architect)
Simona Castelli, Kate O'Daly, Ciara Reddy, Paul O'Brien
项目经理：Kerin Contract Management
结构工程师和土木工程师：PUNCH Consulting Engineers
机械工程师和电气工程师：Don O'Malley & Partners
承包商：PJ Hegarty and Sons
甲方：Plassey Campus Developments
工料测量师：Nolan Ryan Tweed
安全与健康管理：Willis Consulting
消防和通道管理：G. Sexton & Partners
面积：Medical School _ 4300m² /
Student Housing _ 3,600m² / Pergola _ 180m² / Piazza _ 1.2ha
竣工时间：2012.12
摄影师：©Denis Gilbert(courtesy of the architect)

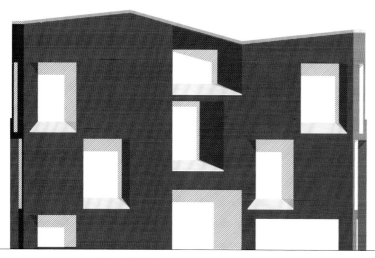

南立面_学生公寓 south elevation_student housing

明暗学习区 light and shade study

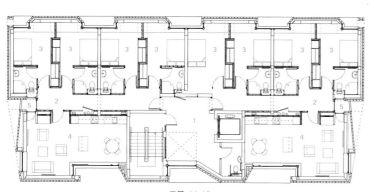

三层 third floor

1 公共区
2 走廊
3 学习卧室
4 厨房/起居空间
5 阳台
6 入口前院
7 拱道

1. common area
2. corridor
3. study bedroom
4. kitchen/living room
5. balcony
6. entrance forecourt
7. archway

C-C' 剖面图 section C-C'

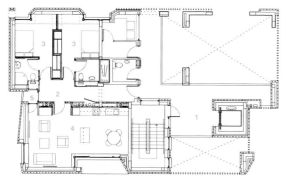

二层 second floor

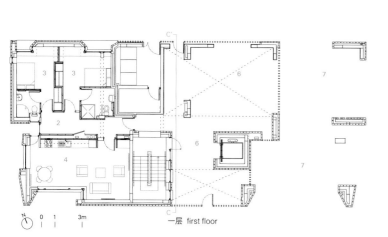

一层 first floor

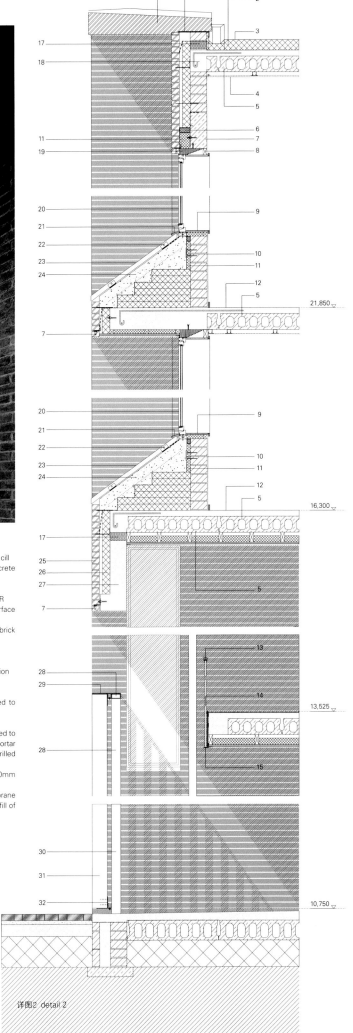

1. artificial slate cavity closer
2. 130mm kingspan thermal roof TR27 insulation board
3. bituminous roofing membrane applied in two layers
4. 12.5mm plasterboard with 2 layers of skim coat fixed to the underside of concealed grid suspended ceiling system
5. 200mm precast concrete unit
6. firestop
7. stainless steel concealed cavity relieving angle
8. stainless steel expanded metal lathe
9. 18mm hardwood window board with 18 x 40mm deephardwood window trim, with rebate as indicated, fixed to treated softwood grounds with packed mineral wool insulation
10. abbey slot & wall tie
11. mineral wool insulation
12. 75mm concrete structural screed
13. 45 x 50mm stained FSC certified iroko handrail screwed and glued to 30 x 11mm ms flat
14. 16mm ms bar welded to handrails
15. 42 x 8 mm MS flat welded to plate to take MS bars
16. cast in-situ concrete coping with sandblasted finish
17. foamglass perinsul SL insulation blocks
18. monarflex stepped DPC bonded to the DPM/radon barrier at ground level & bonded to rc subcill at window cill
19. raised access floor tiles on adjustable pedestals, linoleum floor finish, airtight membrane
20. timber frame (RAL 7016), with double glazed window or solid insulated panel
21. powder coated (RAL 7016) aluminium profile window cill
22. stainless steel bracket anchored to reinforced concrete sub-cill and dowel-fixed to limestone cill
23. limestone cill
24. reinforced concrete sub-Cill cast against preprufe 300R bituminous water proof paint to be applied to top surface of concrete sub-cill
25. stainless steel concealed relieving angle to carry brick outer leaf and suspend one course of brick below
26. 100 mm kingspan kool therm K8 insulation
27. reinforced concrete insitu beam/lintel/column
28. 60mm x 120mm painted galvanised steel hollow section
29. 10mm painted galvanised steel plate
30. vertical jamb dpc
31. 75mm x 200mm solid iroko hardwood posts screwed to steel angle with countersunk screws
32. 50mm x 150mm painted galvanised steel angle
33. limestone seat 600 x 150 with sloped top surface, drilled to receive stainless steel dowel fixing and placed on mortar bed on rc wall, rc wall with bush-hammered finish, drilled to receive stainless steel dowel fixing.
34. 104 stretcher bond random mix concrete setts on 50mm sharp sand on well compacted sub base
35. 105 land drain constructed of weather proof membrane retaining; geotextile layer, sulphate free granular backfill of 40mm clean crushed stone, 150mm diameter half perforated pipe, and 150mm pea gravel
36. 

详图2 detail 2

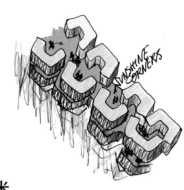

# 维也纳对外经济贸易大学的法学院和行政大楼
CRAB Studio

经过一系列共五场同时进行的设计竞赛后，维也纳对外经济贸易大学将在Prater公园的边上，即靠近标志性的Prater摩天轮处建造一个全新的校园。CRAB工作室赢得了在校园的西南侧设计建造法学院和中央行政大楼的殊荣。

由BUS建筑事务所负责的校园的新总体规划包括修建一系列中央露台，CRAB工作室要设计的建筑从露台延伸出去，以包围他们自己建造的露台、通道和周边环境。

主研究大楼的规划和布局是基于设计师丰富的大学生活体验和对于其独特的学术交流价值所产生的信念的。换句话说，应该为学者教授、研究人员、学生或来访者多提供一些可供他们放松休息、一起聊天或集思广益的空间、随机地点或条件。

设计中人们对部门和分部的追求成为设计师精心设计的"卷曲"或"包覆"空间的灵感。这样，这些用于大家非正式聚会的"小场所"与必要的办公室和研究室相得益彰，琴瑟共鸣。因为该建筑的布局沿着各种活动区域而设计，所以这种正式的和非正式的空间的交织，即所定义的外围空间的设计理念不仅应用于内部空间设计，也应用于阳台、平台和院子这些外部空间的设计。

该建筑设计的焦点是露台式庭院所覆盖的法学院图书馆。图书馆内设计了专门用于延续研讨习俗的走廊，向上抬起，形成瞭望台。

建筑的外表可以看作是一个外围护结构，充满诗情画意，可以毫不费力地交织和改变方向。建筑的外表颜色从下向上由深到浅，一层是深红色的，逐渐向上变成橘色和奶油色，最高一层是白色。建筑的过滤层为木板条，木板条的材质是西伯利亚落叶松（这种木材随着时间的推移会变白，极为坚固耐用）。这些木板条不仅能够遮挡阳光，同时从建筑的角度来说，让人联想到Prater公园森林中的树木屏障。

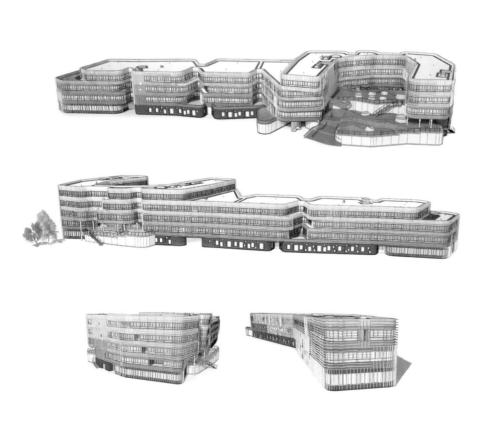

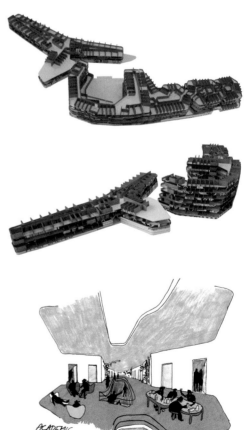

项目名称：Departments of Law and Central Administration, Vienna University
地点：Vienna, Austria
建筑师：CRAB Studio
项目团队：Peter Cook, Gavin Robotham, Mark Bagguley, Stefan Lengen, Theresa Heinen
甲方：Wirtschafts University
建造面积：20,000m²
造价：EUR 29,000,000
竞赛时间：2008
施工时间：2002—2003
竣工时间：2013
摄影师：©Ronald Kreimel(courtesy of the architect)

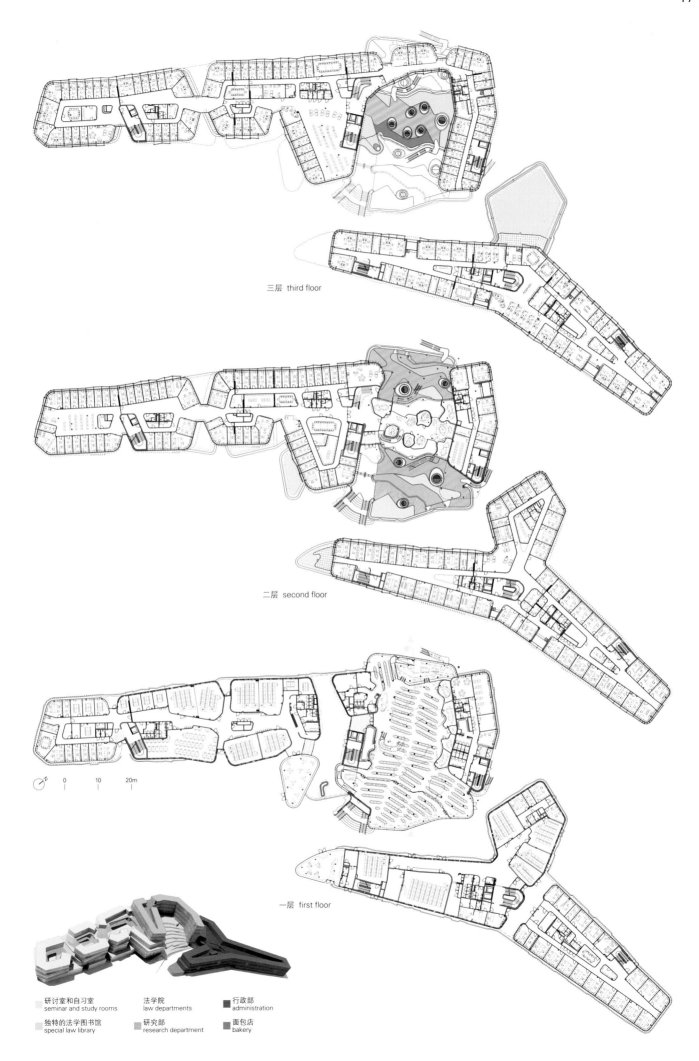

## Vienna University's Law and Administration Building

Following a series of five concurrent competitions, Vienna's University of Economics and Business will create a completely new campus on the side of the Prater Park – close to the heroic Prater Wheel. CRAB Studio won the competition for the pair of buildings on the southwest flank of the campus that houses the Law faculties and the central administration.

The master plan by BUS Architektur establishes a series of central terraces, from which CRAB's buildings proceed to wrap around their own set of terraces, passages and enclaves.

The philosophy behind the planning and the configuration of the main study building emerges from the authors' considerable experience of university life and belief in the value of extra-seminal exchange. In other words – an acknowledgement of the value of spaces, incidental locations or conditions in which academics, researchers, students or visitors will start to unwind, chat or speculate together.

The pursuit of the departmental and sub-departments within acted as an inspiration for the deliberate "curling" or "wrapping" of the plan, thus creating a series of "pockets" for informal gatherings which are orchestrated together with necessary runs of offices and research rooms. As the building has developed along through the various activity zones, this attitude towards the interplay of the formal and the informal – the defined and the peripheral concept has been applied to both internal spaces and the outside balconies, decks and courts.

The focus of the building is the Law Library which is covered by a terraced courtyard. Special study galleries sitting within the library will rise up as "lookouts" towards the terraces and thus continue the tradition of the "scholar's perch".

The wrapping of the building is seen as a lyrical envelope that can weave and change direction effortlessly. As these climb upwards they will change colour from dark red at the ground level, through orange and cream to white at the highest level. A filter layer of slats – cut from Siberian Larch (a timber that bleaches over time and has a tremendous durability) will enable shading from the sun and architecturally establish a link to the timber screening of the Prater woods.

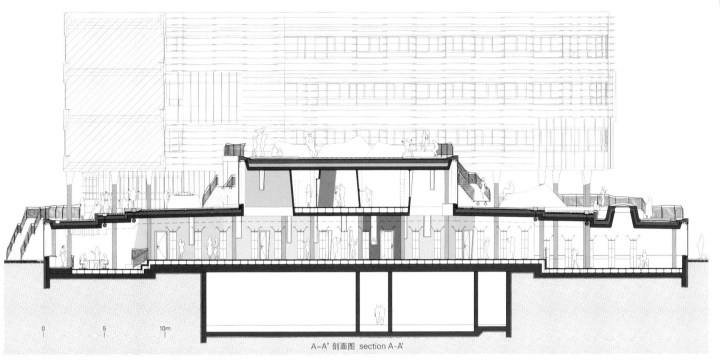

A-A' 剖面图 section A-A'

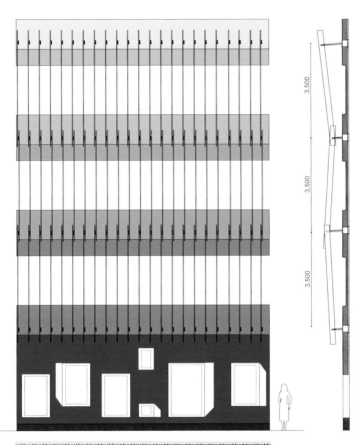

立面的立面图和剖面图 facade elevation and section

vertical lamella holder
Vertical brackets are always mounted on concrete.

type 1. the minimum length of bracket, 300mm
type 2. the maximum length of bracket, 700mm
type 3. minimum and maximum overlapping, 300mm + 700mm
type 4. minimum and minimum overlapping, 300mm + 400mm

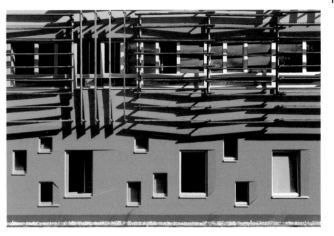

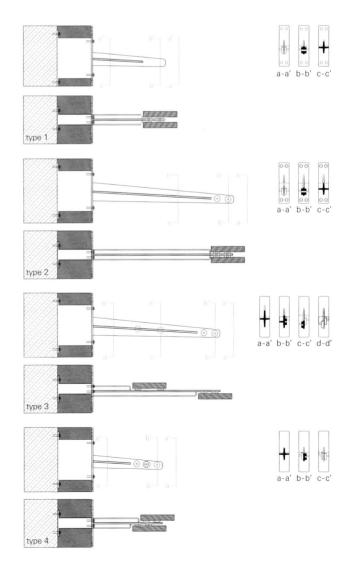

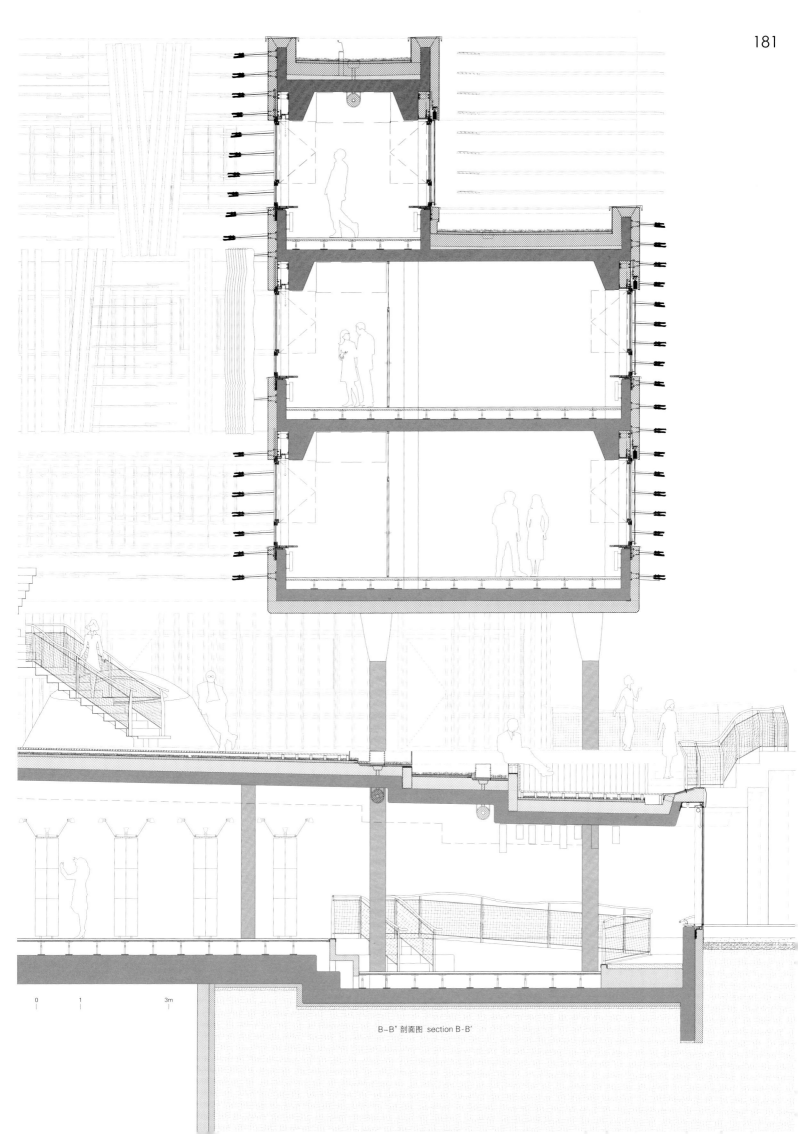

B-B' 剖面图 section B-B'

## >>150
### Taller Básico de Arquitectura
Javier Pérez Herreras was born in Spain in 1967. Is a Ph.D. in Architecture from the University of Navarre. Is also a professor at the University of Zaragoza and visiting professor of IE University. Develops research activity at different universities which is always related to his professional activity.
Javier Quintana de Uña was born in Spain in 1964. Holds a master's degree in Advanced Architectural Design from Columbia University, and has a Ph.D. in Architecture from the University of Navarra. Is the country representative in Spain of the Council on Tall Buildings and Urban Habitat from Chicago.

## >>138
### Tadao Ando Architect & Associates
Tadao Ando was born in Japan in 1941. Is one of the most renowned contemporary Japanese architects. Characteristics of his work contain large expanses of unadorned architectural concrete walls combined with wooden or stone floors and large windows. Active natural elements, like sun, rain, and wind are a distinctive inclusion to his style. Established Tadao Ando Architect & Associates in 1969 and was awarded the Annual Prize of Architectural Institute of Japan for "Row House in Sumiyoshi" in 1979. Has received many awards including Académie d'Architecture in 1989, The Pritzker Architecture Prize in 1995, Gold Medal of the American Institute of Architects in 2002, and Gold Medal of Union Internationale des Architectes in 2005.

## >>158
**Grafton Architects**
Was established in 1978 by two directors Yvonne Farrell and Shelley McNamara. They are both graduates of University College Dublin, and International Honorary Fellows of the RIBA at the same time. Co-directors Gerard Carty and Philippe O'Sullivan have been working in the practice for 20 years. Core members are Simona Castelli, Kieran O'Brien, Matt McCullagh, James Rossa O'Hare, Donal O'Herlihy, David Healy, Joanne Lyons, Ivan O'Connell and Edwin Jebb. They are winners of numerous awards including the World Building of the Year Award 2008 for their building for the Luigi Bocconi University–School of economics in Milan, Italy.

## >>168
**CRAB Studio**
Peter Cook right was born in Southend-on Sea in 1936. Studied at Art University Bournemouth and AA(Architectural Association School of Architecture) in London. Was knighted by the Queen in 2007. Is a Royal Academician and a Commander of the Order of Arts and Letters and a Senior Fellow of the Royal College of Art, London.
Gavin Robotham left was born in North Walsham in 1969. Studied at Bartlett School of Architecture and Harvard Graduate School of Design. Is the director and design team leader of CRAB Studio and continues to lecture design and acts as a consultant.

## >>28
**AR Design Studio**
Andy Ramus, a director of AR Design Studio was born in London and was trained at Plymouth School of Architecture under the eminent professor Adrian Gale. In 2000, he completed his education at the AA(Architectural Association School of Architecture) and established his own practice in 2001. With a comprehensive knowledge of planning, he has built a stunning portfolio of completed residential projects, from existing home extensions and re-modeling to new-builds and multi-plot developments. Received numerous awards including the Best of Houzz Award 2013, the RIBA Downland Award 2012, the Federation of Master Builders Regional Award 2012 and the Daily Telegraph Small House of the Year Award 2008.

## >>94
**Manuel Ocaña**
Received a master's degree in Science of Architecture in 1992. Studied at ETSAM where he currently lectures. Was a visiting lecturer of Materiality at Barcelona Institute of Architecture and now directs an on-line design studio at the IE School of Architecture & Design. Has participated in diverse international exhibitions and lectured worldwide. His ideas and built work have been awarded several prizes. His work has been widely published in El Croquis, Bauwelt, Arquitectura Viva, Metalocus, Volume and so on.

**Aldo Vanini**
Practices in the fields of architecture and planning. Had many of his works published in various qualified international magazines. Is a member of regional and local government boards, involved in architectural and planning researches. One of his most important research interests is the conversion of abandoned mining sites in Sardini.

## >>82
**Cannatà & Fernandes Arquitectos**
Michele Cannatà is an Italian architect who was born in 1952. Received his Ph.D. in "Composizione Architettonica e Progettazione Urbana" from the University of Chieti-Pescara, Italy in 2009.
Fátima Fernandes is a Portuguese architect who was born in 1961. In 1986, graduated from the fine arts faculty of the University of Porto. She has been working as a director of ESAP's Architectural Design school since December 2012. They have been working together since 1984 and founded Cannatà & Fernandes Arquitectos in 2000.

**Tom Van Malderen**
His activities stretch from the traditional architectural practice to the field of architectural theory which he explores through writing, installations and lectures. After obtaining a master in Architecture at LUCA, Brussels(1997) he worked for Atelier Lucien Kroll in Belgium and in different positions at architecture project, both in the UK and Malta. Lectured at the University of Aix-en-Provence in France and the Canterbury University College of Creative Arts in the UK. Contributes to several magazines and publications, and sits on the board of the NGO Kinemastik for the promotion of short film.

**Paula Melâneo**
Is an architect based in Lisbon. Graduated from the Lisbon Technical University in 1999 and received a master of science in Multimedia-Hypermedia from the cole Supérieure de Beaux-Arts de Paris in 2003. Besides the architecture practice, she focused on her professional activity in the editorial field, writing critics and articles specialized in architecture. Since 2001, she has been part of the editorial board of the portuguese magazine "arqa –Architecture and Art" and the editorial coordinator for the magazine since 2010. Has been a writer for several international magazines such as FRAME and AMC. Participated in the Architecture and Design Biennale EXD'1 as editor, part of the Experimentadesign team.

>>36
**Hironaka Ogawa & Associates**
Hironaka Ogawa was born in Kagawa in 1975, and completed a master course at the Department of Architecture in Nihon University in 2000. After graduation, he worked at Kengo Kuma & Associates for 5 years and established Hironaka Ogawa & Associates in 2005. Received several awards in various competitions including Design For Asia Award 2006 and NASHOP Lighting Awards 2007. Currently lectures at Toyo University, Kagawa University, and Nihon University. Is one of the young architects who suggest new possibilities in architecture by uniting the materials, the new technologies, and aesthetic sophistication.

>>106
**Carlos de Riaño Lozano**
Received a degree in Architecture from the Superior Technical School of Architecture of Madrid. Opened his practice at Madrid, Spain in 1980 and obtained many awards for his completed projects throughout his career in several national and international competitions. It includes C.O.A.M. Award, Europa Nostra Award, Madrid City Council Award, VETECO Award and Sánchez Esteve Award. His work has been published in a variety of Spanish and international publications. Has been a lecturer and visiting professor at many Spanish universities and institutions. Is currently based in Madrid.

## >>66
**HLPS Arquitectos**
Jonathan Holmes, Carolina Portugueis, Osvaldo Spichiger, and Martin Labbé are leading the practice. They have participated in public competitions and have awarded 1st prize in the competition for the Metro Station Cal y canto and the Parque Cultural Valparaiso in the former jail complex. Their work have been exhibited in various architecture Biennial in Chile and have been published in diverse specialized magazines such as ARQ, CA, Casabella, Architectural Review, Summa and Arquine.

## >>48
**BIG**
Was founded in 2005 by Bjarke Ingels. Is an architectural office currently involved in a large number of projects throughout Europe, Asia and North America. Their architecture emerges out of a careful analysis of how contemporary life constantly evolves and changes, not least due to the influence of multicultural exchange, global economic flows and communication technologies that together require new ways of architectural and urban organization.

## >>118
**Jaime J. Ferrer Forés**
Was born in 1975 in Palma de Mallorca, Spain. Founded Ferrer Forés Architects after receiving Diploma in Architecture(2000) and Ph.D.(2006) from ETSAB. His office has won various awards for numerous works. Is teaching at the Architectural Design Department in ETSAB of the Polytechnic University of Cataionia. Is the author of the monograph of the architect Jørn Utzon who designed Sydney Opera House. His works were published by Gustavo Gili in 2006 and exhibited at the 13th Venice Architecture Biennale in 2012.

## >>8
### AFF Architekten

Was founded in 1999 by Martin Fröhlich[left], Sven Fröhlich[right] and Torsten Lockl, graduates of the Bauhaus University in Weimar. The firm is the result of team work and they received awards and high regards not only in Germany but also around the world.

Martin Fröhlich was born in Magdeburg in 1968. Studied architecture at the Bauhaus University of Weimar for 5 years from 1989. After graduation, he lectured at the Bauhaus University of Weimar, Berlin University of the Arts and Swoss Federal Institute of Technology Lausanne.

Sven Fröhlich was born in Magdeburg in 1974. Studied architecture for 6 years and visual communications for 5 years at the Bauhaus University of Weimar.

©Dawin Meckel

## >>18
### Camponovo Baumgartner Architekten

Marianne Julia Baumgartner[left] and Luca Camponovo[right] both graduated from the ETH Zürich in 2009 and founded Camponovo Baumgartner Architekten in 2010 in Zürich.

Marianne Julia Baumgartner was born in Bern in 1984. She worked for several offices in Lausanne, Bern and Barcelona after graduation. She was teaching assistant at the Chair of J.Ll.Mateo at ETH Zürich between 2010 and 2012.

Luca Camponovo was born in Burgdorf in 1980. He worked for several offices in Zürich, including KCAP, Burkhard Meyer and mainly Müller Sigrist Architekten after graduation.

C3,Issue 2014.1
All Rights Reserved. Authorized translation from the Korean-English language edition published by C3 Publishing Co., Seoul.

© 2014大连理工大学出版社
著作权合同登记06-2014年第51号

**版权所有·侵权必究**

### 图书在版编目(CIP)数据

记忆的住居 ：汉英对照 / 韩国C3出版公社编 ；朱黛娜等译. — 大连：大连理工大学出版社，2014.4
(C3建筑立场系列丛书)
书名原文：C3:memory
ISBN 978-7-5611-9027-2

Ⅰ. ①记… Ⅱ. ①韩… ②朱… Ⅲ. ①住宅－建筑设计－汉、英 Ⅳ. ①TU241

中国版本图书馆CIP数据核字(2014)第065073号

出版发行：大连理工大学出版社
（地址：大连市软件园路80号　邮编：116023）
印　　刷：上海锦良印刷厂
幅面尺寸：225mm×300mm
印　　张：11.75
出版时间：2014年4月第1版
印刷时间：2014年4月第1次印刷
出 版 人：金英伟
统　　筹：房　磊
责任编辑：张昕焱
封面设计：王志峰
责任校对：赵姗姗

书　　号：ISBN 978-7-5611-9027-2
定　　价：228.00元

发　行：0411-84708842
传　真：0411-84701466
E-mail：12282980@qq.com
URL：http://www.dutp.cn